An Ocean of Oil

A Century of Political Struggle Over Petroleum Off The California Coast

AF531034

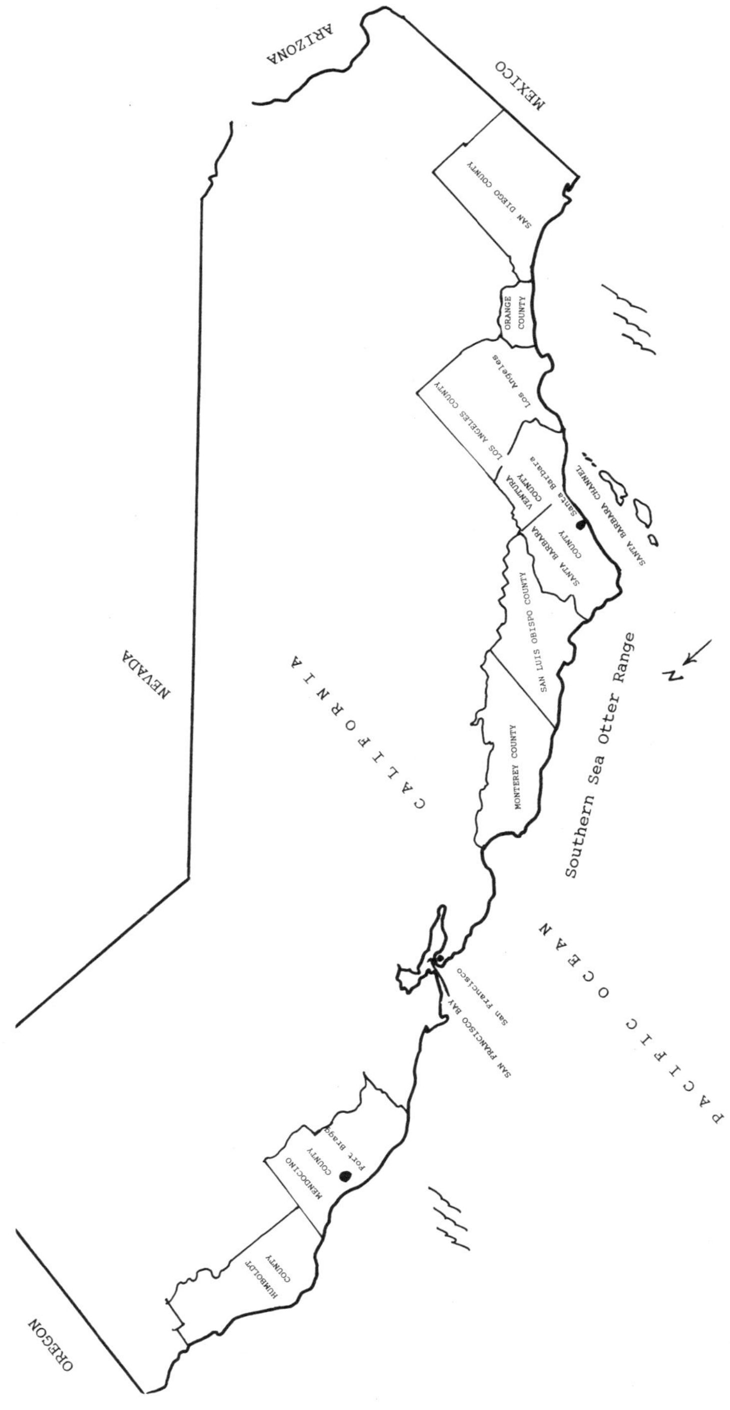

California Coast

An Ocean of Oil

A Century of Political Struggle Over Petroleum Off The California Coast

Robert Sollen

The Denali Press

Denali, derived from the Koyukon name *Deenaalee*, is the native name for Mount McKinley. Mount McKinley, the highest mountain on the North American continent, is located in Denali National Park and Preserve. The lowlands surrounding this majestic mountain provide a diverse wildlife habitat for a variety of animals, including grizzly bears, wolves, caribou and moose.

Copyright © 1998 by The Denali Press

Published by The Denali Press
Post Office Box 021535
Juneau, Alaska USA 99802-1535
Phone: (907) 586-6014 Fax: (907) 463-6780
E-mail: denalipr@alaska.net http://www.alaska.net/~denalipr

All rights reserved. No part of this publication may be reproduced, stored in a retrieval system, or transmitted in any form or by any means, electronic, mechanical, photocopying, recording or otherwise, without the prior written permission of The Denali Press.

Library of Congress Cataloging-in-Publication Data

Sollen, Robert
An ocean of oil : a century of political struggle over petroleum off the California coast / Robert Sollen ; with a foreword by Roderick Nash.
p. cm.
Includes bibliographical references and index.
ISBN 0-938737-34-1 paper : alk. paper) : $35.00
1. Offshore oil industry--California--History. 2. Offshore oil industry--Environmental aspects--California--History. 3. Petroleum in submerged lands--California--History. 4. Offshore oil industry--Government policy--California--History. I. Title.
HD9567.C2S65 1998
338.2 ' 7282 ' 09794--dc21 97-33453
CIP

∞ The paper used in this publication meets the minimum requirements of the American National Standard for Information Sciences—Permanence of Paper for Printed Library Materials, ANSI Z39.48. This book is printed on recycled paper, using soy based ink.

For Tomi

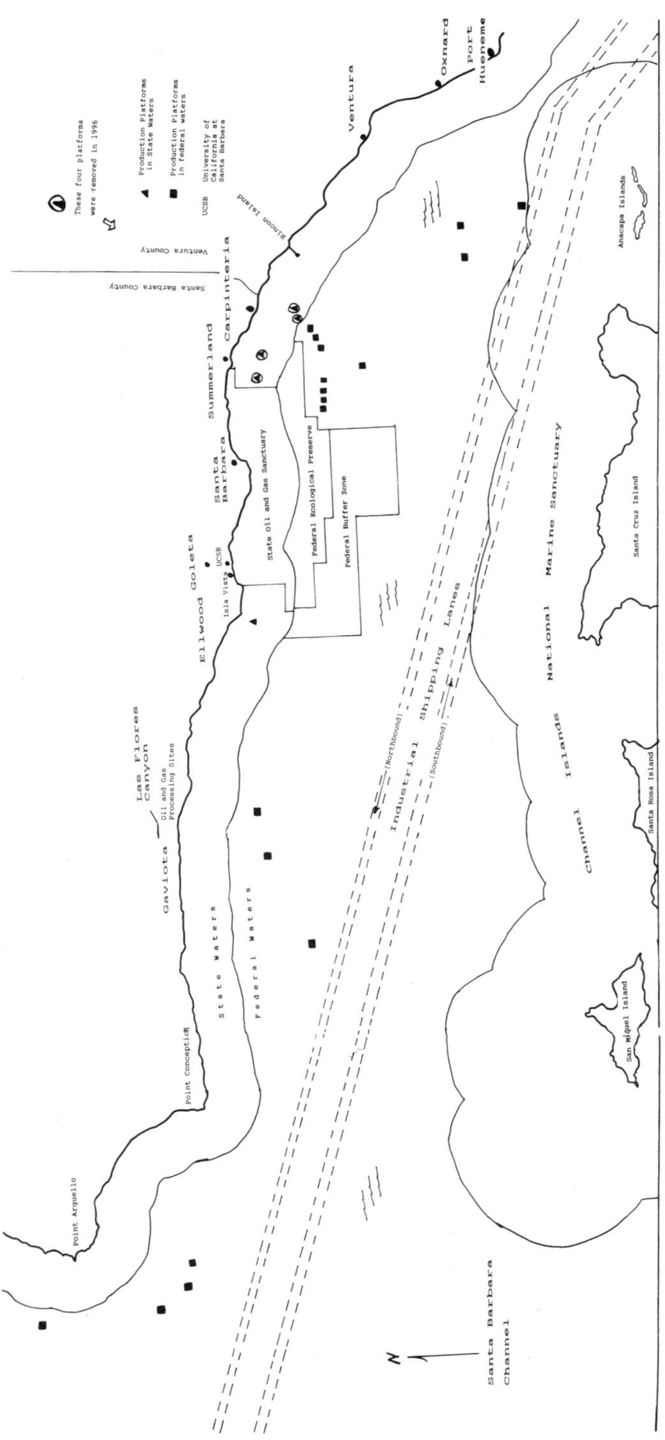

Santa Barbara Channel

Contents

Foreword

There has been oil on the water frequently in the history of California offshore oil production, but the timing of the Santa Barbara Oil Spill of 1969 lent special significance. The 1960s in general was a time of questioning of basic American values. Some thought in terms of a "counterculture." At its core was the belief that bigger was not always better, that government was not always benevolent and that technology might not lead directly to the good, sustainable life.

Rachel Carson opened the 1960s, and created modern American environmentalism, with her blockbuster book *Silent Spring*. It raised the problem of the irony of too much control over nature. Sure, the miracle pesticides like DDT could kill undesirable insects, but as the poison worked through the food chains, what else died? Interestingly, Carson's Exhibit A was the same form of life that became critical after the 1969 blowout: birds. Her's were belly up ("silent" as the title implied) from ingestion of pesticide-contaminated insects. Santa Barbara's were soaked in oil, flapping out their final hours on blackened beaches.

As symbols of an environment blighted by technological and political blunders, film and photographs of oil-covered birds circled the globe. The impact was comparable to images of the beatings and hosings of black Americans that made front pages a decade before the Santa Barbara spill and fueled the civil rights movement. Here were new victims of oppression and a new cause for liberal idealists. The events of Santa Barbara helped environmentalism join civil rights and women's liberation on the plate of American reform.

There could not have been a more dramatic place to test the benefits and costs of technology with respect to the environment and quality of life than on California's legendary South Coast. Indeed, it is arguable that if the 1969 spill had occurred off, say, a remote marshland in Louisiana or the frozen Alaskan tundra, there would have been relatively little protest. But southern California, and Santa Barbara in particular, was a recreational paradise. Tourism had always thrived here, and, as Bob Sollen shows, local misgivings antedated the 1960s. Summerland was said to be "aslime with oil leakages" in 1911; as heavy a word as "rape" appeared in connection with the opening of the Ellwood oil field in the late 1920s. For years after World War Two seismic test explosions rocked the Santa Barbara Channel and disturbed marine life.

While a paradise lost, or at least compromised, was vital to the significance of the Santa Barbara Oil Spill, so was the presence in this California community of many liberal, angry intellectuals. Centered at the University of California Santa Barbara, where I taught environmental history, and at the *Santa Barbara News-Press*, for which Bob Sollen wrote, public reaction to the spill was intense and emotional. My own might be a case in point. The news that a vast layer of oil was moving toward the South Coast early in February 1969 was shocking in the extreme. As usual the Channel Islands framed the southern horizon, palms swayed and beaches gleamed white sand, but a menace was at the door. It seemed to some of us like one of those old science fiction nightmare films featuring alien blobs or ooze. I recall taking my eight-year-old daughter down to the beach to see what happened. People walked around stunned; some knelt among the dead birds and blackened kelp and wept. It was difficult to explain to a child why this had happened. Playing the objective professor, I reminded her that in the big picture, we were part of the problem. We had, after all, used a gasoline engine to reach the beach and see the black petroleum tide.

Later I wrote *The Santa Barbara Declaration of Environmental Rights*.* I did it at sea, on Barry Schuyler's boat *Wishbone*. I had a copy of Jefferson's *Declaration of Independence* beside the wheel, and in view of the oil spill it seemed appropriate to think about the extension of natural rights to the rights of nature. I could not suppress the thought that in regard to technology and environmental impact our species was at an historical crossroads. One path led to continued exploitation, unlimited growth and the collapse of the planet's life-support capabilities. The other might take us to peaceful coexistence with the larger life community and a sustainable future. Just as Jefferson's generation had ethical choices to make, so does ours. Santa Barbara helped bring them into focus.

Dr. Roderick Nash

*See page 67

Preface

The world's first offshore oil wells drilled in 1896 off Summerland, California, opened a century of industrial progress, environmental degradation, and confrontation between coastal communities and the industry, each with its partisan supporters.

Many more wells have been drilled in the Gulf of Mexico than off California. But the Gulf wells are far offshore and dispersed. Off the West Coast they are concentrated near urban areas, mostly near shore in the Santa Barbara Channel between the mainland and the Channel Islands, in clear view from the coastal communities.

The California offshore wells, particularly off Santa Barbara and Ventura counties, have created a series of booms and busts, and with each cycle a natural resource catastrophe. Local economies sometimes benefitted from the booms, but the busts always placed any benefits in question. California taxpayers are still spending millions of dollars to remove coastal debris from a century ago.

Although the 1969 oil well blowout in the Santa Barbara Channel is often cited as the catalyst for the environmental movement, protests against offshore oil did not start with that event. For a century, Santa Barbarans and other Californians have protested coastal oil activity as a messy business.

As a reporter for the *Santa Barbara News-Press*, I provided coverage beginning in February 1968 when the federal government conducted its first major offshore oil lease sale in the channel. The newspaper assigned me to report on offshore oil developments. For seventeen years I covered these contentious activities. In the process I accumulated an extensive California offshore oil history. When I retired in mid-1985 I became the Sierra Club offshore oil policy coordinator for the Santa Barbara area. In January 1989 I was appointed to the county Planning Commission which perpetually wrestled with offshore oil disputes. After two and one-half years I returned to work as a Sierra Club volunteer.

The hectic one hundred year struggle over offshore oil shifted, often chaotically, from the courthouse to the White House, from Sacramento to Washington, from City Hall to Congress, and the

United States Supreme Court. It created transcontinental commuters of local and Washington delegations. Local governments, the state legislature, Congress and the Interior Department tried to control an industry that always seemed to be a couple jumps ahead of adequate regulation. Coastal communities implored presidents and Interior Department secretaries who often cared less about these communities and their concerns than of the industry and the federal revenue the industry created.

Until now this story has been told only in fragments. This book brings it together, not as a technical account, but as an informal sociopolitical history, emphasizing the repercussions of the industry on California coastal communities. This is that story.

ACKNOWLEDGMENTS

More people and agencies helped in the preparation of this book than can be conveniently listed . But you know who you are, and I am grateful. Any errors, of course, are mine alone.

Among those who read parts of the manuscript and helped improve it were Robert Easton, author of *Black Tide*; Paul Veblen, former executive editor of the *Santa Barbara News-Press;* Amy Wiltse, environmental studies graduate student at the University of California at Santa Barbara; Albert McCurdy of the Santa Barbara County Planning and Development Department; Al Reynolds, former director of the Santa Barbara County's Office of Environmental Quality; Linda Krop, senior counsel for the Santa Barbara Environmental Defense Center; and Jack Hundley, former regional executive for Atlantic Richfield Company.

Staff members of many public agencies assisted with research: the counties of Santa Barbara and Orange, especially the Santa Barbara County Energy Division and the Santa Barbara County Association of Governments; state agencies, including the State Lands Commission, Coastal Commission, and the Division of Oil and Gas. Primary among federal agencies was the Minerals Management Service.

The *Santa Barbara News-Press* was generous in permitting the use of its photographs. The *San Clemente Sun-Post* granted permission to use one of its photographs.

Among nongovernmental agencies providing information were the Huntington Beach Historical Museum, Santa Barbara Historical Museum, the California Maritime Academy, Save Our Shores of Santa Cruz, the Carpinteria Museum of History, the Unocal Oil Museum in Santa Paula, and the Ventura County Historical Museum. All staff members were eager to help.

Also helpful were chronologies prepared by the staff of the California Coastal Commission; by Barry Cappello, former Santa Barbara city attorney; and by Ruth Corwin and Reed Holderman.

Information from many other public and nongovernmental sources is used throughout, and is cited in the text or in endnotes. Books referred to are listed in the Select Bibliography.

Newspaper reports are ordinarily not primary documents and therefore seldom the best sources for historical research. When used here they were usually either corroborated or were my own *Santa Barbara News-Press* reports which were written from primary sources, including public documents, public testimony, field trips, interviews, etc. Beyond that, news reports were used when there was good reason to believe they were reliable and when primary sources were not available, or occasionally simply to relate an attitude of the news media.

I did little interviewing for interpretations of events for this book. I relied largely on documents and my notes prepared at the time. With few exceptions, interviews and requests were for factual information and not for recollected descriptions or interpretations of events. The Carpinteria Museum of History has a valuable collection of oral histories of the early days of offshore oil. Some "old-timers" wrote books. Some local historians wrote books and articles.

I conducted two long interviews for descriptions of events I did not witness. I talked with William Master and Kathy Milway about events leading to the federal Clean Air Act amendments of 1990. They were at that time members of the Santa Barbara County Air Pollution Control District who lobbied for amendments favorable to the county. Robert Kallman, a former director of the Minerals Management Service, was interviewed for a description of the 1988 federal hearings in Northern California on proposals for oil leasing off those shores. He was one of several sources used to describe this event.

The reference material used in the preparation of this book is now in the Special Collections Department of the Davidson Library at the University of California at Santa Barbara.

Chapter One

From the Beach, a Short Step Into the Ocean

The world's oil industry is young and is already approaching its peak. Petroleum itself is a recent addition to the earth. It was not present at the creation 4.6 billion years ago. It took millions of years to evolve, and the human race is consuming it at the rate of sixty to seventy million barrels a day. The evolution of oil did not even start until nature produced a profusion of living creatures, the decomposed remains of which are the raw materials of petroleum. That process started about six hundred million years ago, or four billion years after the creation of the earth.

A few tens of thousands of years ago indigenous Americans found many uses for oil that oozed through the earth's surface. Nineteenth century European immigrants went after it more aggressively, gouging oil tar from cliffs, bluffs, and flatlands to refine for heating, lighting, road paving, lubricating, roof sealing, and other limited purposes. Petroleum refining, then, preceded oil well drilling and automobiles.

Not until Edwin L. Drake drilled the first successful oil well in Titusville, Pennsylvania, in 1859 was there—for better or for worse—an oil "industry." Californians had already been drilling for four years in several places, but it was not until the 1870s that the first successful wells were drilled in California.

Like most modern science, petroleum geology is new. But it did not require a petroleum geologist to tell early Californians that the West Coast was oil country. Since before recorded history, oil, tar and natural gas have seeped like languid artisan wells through the surface of the earth and the seafloor. Onshore seeps were found along the California coast, and many more inland. But most of the seeps were offshore what is now Santa Barbara County. That is where the California coast abruptly shifts to an east-west orientation for forty-five miles north of the four northern Santa Barbara Channel Islands which also run east-west. Between these islands and the mainland is the Santa Barbara Channel.

Free Oil and Gas

Petroleum evolved from plant and animal matter deposited in the sediments of inland seas and coastal marine basins. Organic material that was buried under sediments before it could be oxidized decomposed into hydrocarbons, the stuff of oil and gas. Petroleum did not form in pools, but in droplets between grains of subsurface sand and rock. Lighter than water, which is also present, oil rises between sand and rock particles until it meets an impervious barrier such as a fault trap or an anticline. An anticline is a geologic formation warped upward under pressure to form a dome which, if impervious, can trap oil that migrates upward.

Some of the migrating oil and gas does not stay trapped; it rises to the surface through the sea floor or onshore formations. This happened where oil-bearing formations were warped upward toward the surface, and then were eroded or fractured, allowing the oil and gas to escape to the surface. Most seeps show no signs of running dry, as oil and gas from deep down or far out continue to migrate into areas where they can escape to the land surface or into the ocean. Can the seep oil deposits be recovered through oil wells? Not much of the seep oil would come up through a well when it has so many other places—paths of lesser resistance—to emerge. One experiment in trapping seeping gas, however, has produced remarkable results. More of that later.

Oil and gas seeps, then, are not unusual, but those off the Santa Barbara County shore are possibly the world's most profuse. They are largely in the Carpinteria-Summerland region ten to fifteen miles east of Santa Barbara, and from Coal Oil Point to Point Conception, ten to forty miles west of the city. Seep statistics are unreliable, useful only to dramatize the colossal volume of hemorrhaging petroleum. Seep study has been fragmentary, and each new research paper emphasizes how much is not known. Much of the bibliography is a series of papers citing a few cursory field studies.

No one knows, for instance, how long the Santa Barbara offshore seeps have existed. At least ten thousand years and perhaps during all of later Pleistocene time, writes Peter J. Fischer in a 1976 paper prepared for the Federal Energy Administration.[1] Some researchers have tried to count the number of seeps in the area, a futile exercise since the number varies wildly even from day to day. And they keep moving around. One 1977 state report indicates two thousand seeps have been identified along the Santa Barbara County coastline, and that six hundred sixty-five more may be "inferred." The rate of flow is also elusive. How fast oil and gas come up defies calculation because there is no reliable technique for it, and the volume is highly erratic. Several hundred barrels per day is a generally-accepted guess, averaging perhaps five hundred.

Where does all this oil go? Many places. When an oil blob reaches the surface of the ocean, the lighter (volatile) elements evaporate, entering (and polluting) the atmosphere. The remaining oil is heavier, and some of it sinks. Some of it drifts to oceanfloor canyons and settles there. In choppy seas, the oil is mixed into an emulsion, like chocolate syrup, and some of it comes ashore on the wind and the running sea. In summer, when seas are calm, the oil does not break up as readily; it accumulates in layers on the ocean surface. When this heavy liquid comes ashore the protest is heard in the offices of the Coast Guard, the oil industry, the newspapers, chambers of commerce, city hall and the courthouse. Many blame the offshore oil industry and the industry replies that the seeps were there long before the oil companies came. The Coast Guard almost always says it is seep oil. There is a difference between industrially-produced petroleum and seep oil, and technically each can be identified. But the process is costly and it is even more difficult to identify the difference when the oil is weathered.

Government, industry and academe have investigated the chance that industrial oil activities induce or increase seepage. Indications are that they do not, but the studies warn that the findings are not conclusive. So the question remains unresolved. Long-time residents and commercial fishers each year insist that there is more oil than ever on the water and the beaches. But more than oil is involved.

Natural gas comes up in bubbles that pop when they reach the surface of the water, and one would think that ends it. But the invisible gas arrives in such volumes that it becomes a serious air polluter. Santa Barbara County's air quality experts said in 1982 that twenty-five percent of the county's reactive hydrocarbons, which turn to photochemical smog, came from offshore gas seeps. The Atlantic Richfield Company (ARCO) developed a plan to trap some of this pollution. ARCO wanted to drill more offshore exploratory wells, and was required to offset the resulting air pollution by reducing pollution elsewhere nearby. The reduction had to equal one hundred twenty percent of the new pollution.

ARCO built two steel pyramid tents, one hundred feet on a side. They were to be lowered over a profuse gas seep area one and one-half miles off Coal Oil Point ten miles west of Santa Barbara where the seeps are perhaps the world's most active. Trapped gas would rise inside to the peak of the pyramids, then pass through hoses to an onshore processing plant and into pipelines to consumers.

To find the busiest gas seeps, Nekton Inc., ARCO's consultants, surveyed the seabed with its sixteen-foot submersible, *Nekton Gamma*. In November 1981, before the steel tents were lowered, *Nekton* took me down two hundred twenty feet to see the gas popping through the ocean floor.

"Like descending in a sea of champagne," said *Nekton* skipper Tim Caffarella as we sank slowly and quietly to the bottom. The incredible statistics on escaping gas became believable as I watched millions of gas bubbles, some big as golf balls, rise through craters in the silt. Marine plants and fish were in profusion. This is where the metal tents would be.

The oil company and the county hoped the structures would trap a half million cubic feet of gas per day. They did not. They trapped *twice* that volume at the outset. And it kept increasing. By 1988, the pyramids collected nearly *one and one-half million* cubic feet daily. After eleven years it was still averaging about a million cubic feet per day. All this in twenty thousand square feet, less than half an acre.

Indigenous, Ingenious

To white settlers and their progeny, petroleum oozing up through the ground and the ocean floor was and is a nuisance. No so to indigenous Americans. The major Chumash villages along the Santa Barbara Channel were built near asphaltum seeps,[2] and the Chumash and other California tribes depended heavily on this material.[3] They regarded the gratuitous petroleum neither as a nuisance nor as a resource to be exploited rapaciously. They simply put it to many good uses and did not fret because there was a lot left over. They used the available oil and tar without ravaging other resources in the process.

Oil tar was used by the Chumash most conspicuously to keep their remarkable plank canoes afloat. Called *tomols*, these ten to thirty foot canoes were made of planks hewn from logs that drifted downcoast from northern timber country. The planks were not overlapped, but placed edge-to-edge, and caulked with a mixture of seep tar and pine pitch. Their expertness amazed the early Europeans, and the Spanish soldiers with the Gaspar de Portola expedition in 1769 named the construction site Carpinteria, Spanish for carpentry or a place where it is practiced.

Despite the skilled craftsmanship, the *tomols* leaked. Someone had to keep bailing with a basket waterproofed with asphalt. The basket was lined with tar that was first pulverized into powder. Hot stones twirled around inside the basket melted the powder and worked it into the willow, rush, tule, cattails or other materials used in basketmaking. Over time the organic plant materials disintegrated, but remnants of petroleum linings exist which display the imprint of the basket patterns.

Petroleum tar not only made *tomols* capable of coastal and inter-island trade, but was itself a commodity of trade. Chumash mariners carried congealed seep oil in balls or cakes to Santa Catalina Island, twenty miles seaward of what is now Los Angeles, and more than fifty miles over water from Carpinteria. In this sense, *tomols* were the world's first commercial marine oil carriers.

Asphalt was good for more than keeping canoes afloat and waterproofing baskets. The Chumash and many other tribes used it in housekeeping, hunting, warfare, art, games, food preparation, and even clothing.

The coastal Chumash and the Yokuts of the inland San Joaquin Valley pulverized seeds and vegetables on stone surfaces. To prevent splattering the food in the process, a bottomless basket was glued to the mortar surface with seep tar, and the basket retained the food being mashed and splattered.[4] As a mastic, oil tar was used to repair broken bowls and stone tools. Stone bowls and other articles were often adorned with shell fragments set in asphalt.

Chumash used asphalt to attach knife blades, arrowheads and spear points to handles and shafts, then often secured by fiber twine or fine sinew wrapped around the joint. Fishhooks of cactus spines, shell, wood and bone, were attached to wooden shanks with tar.

In dress and recreation, solid tar also had a role. Chumash and Yokuts women's skirts of woven tule or sea grass were held down in the wind by asphalt weights pinched onto the bottom. Games were played with dice made of walnut shell halves packed with tar.

"It is clear that the Indians used the asphaltum and heavy oils for symbolic, decorative, and practical purposes," Susan F. Hodgson wrote in a report for the California Division of Oil and Gas.[5] "The Yokuts, Achomawi, Maidu, and Chumash Indians all made use of asphaltum and heavy oils from seeps in various ways."

Oil tar was used over a vast geographic area. It was employed by the Yokuts in the San Joaquin Valley, the Achomawi in the far north of California in Shasta County to the Oregon border, and the Maidu who dwelled in the drainage areas of the Feather and American Rivers, extending eastward from the Sacramento River to the crest of the Sierra Nevada.

Some Indians chewed oil tar. Others, as well as their European successors, advertised petroleum fumes as therapeutic. Occasionally the Chumash molded an oil tar cast for a broken arm or leg.[6]

With a limited population and a reverence for the earth, the indigenous Americans took the resources they needed without depleting or degrading their habitat. Things changed with the arrival of the settlers from Europe. They introduced a population boom, an industrial technology, and a new attitude toward natural resources.

Roofs, Roads, and Lamps

Industrial oil extraction did not wait for the invention of drilling. By the 1850s, asphalt deposits near Goleta Beach west of Santa Barbara were "dynamited into fragments and shipped to San Francisco to pave the streets....The Carpinteria asphalt deposit was exploited on a larger scale. It was first mined in 1857 in an unsuccessful attempt to produce illuminants [fuel for lanterns]. It was later extensively quarried to provide paving materials."[7] Summerland and Carpinteria for a long time furnished all the asphalt for roofs and streets in San Francisco.[8]

The Carpinteria deposits were quarried a short time for distilling lamp oil, and as late as 1920 for paving material. It was open pit mining with all its consequences. David Banks Rogers in 1929 described the scene:

> Today, if one stands at the location of the former asphalt spring, he sees before him, extending from the beach line to the railroad right-of-way and to the right and left for several hundred feet, a yawning crater, devoid of soil and vegetation, except where a few tules cling to the rim of a fetid lagoon at the bottom of the pit. West of this abandoned pit-mine, the activity of many wheels incident to the working of the mine, coupled with periodic winds, has swept the land bare, down to the tough sub-soil.[9]

The pit was the site of a large former Chumash village, later engulfed by the city of Carpinteria.

But there was more than oil and tar in the ground. There was gas. Lots of it. The first gas well was an accident. In 1890 a Summerland resident was digging for hot sulphur water when he got a rush of natural gas instead. It was not wasted. The unanticipated resource was diverted to the lamps of the town's new hotel.

It was soon obvious there was much more gas where that came from. When sandlot baseball games went on until after sunset, Summerland youngsters drove pipes into the ground and ignited the escaping gas. The game went on with light from flames a foot high. When the game was over, players hit the top of each pipe with a board to smother the flame.[10]

"Drilling" Onshore

While there were some uses for petroleum, the market was limited and the oil, tar, and gas on hand constituted a glut. Still, optimistic oil operators in the early 1860s kept at it, and worked out an improvement over hot shovels and open pits for extracting oil. It was called cable tool drilling, which in fact was neither new nor drilling. It was pulverizing, a technique the Chinese used seven thousand years earlier to dig brine water wells. A heavy piece of metal with a chisel-like point was suspended by rope or cable and dropped from a derrick to the earth. Each time it dropped it broke up a little more soil and rock. At intervals the hole was flushed and the slurry was removed in a bailer with a valve in the bottom. It was slow going. Some days in Pennsylvania, Edwin L. Drake, who drilled the world's first oil well, penetrated no more than three feet. But a continent away, in California, prospectors were encouraged.

The first wells were drilled up north in Humboldt County, but they did not produce enough to pay the bills. In Ojai to the south drillers produced a gusher in 1867.

According to a 1988 report from the California Division of Oil and Gas "this was the best well to date and would be considered as the first oil well in California to produce on a commercial basis, except that a record of whether the well produced continually or intermittently is not available."[11] The first "truly" commercial oil well in the state, the report indicates, was drilled in Pico Canyon in Los Angeles County in 1876. And there is a state historical monument there to "prove" it.

The first well in Summerland was drilled the next year, but at two barrels a day it was not celebrated. Other wells in the 1870s did no better. In 1886, however, Henry Lafayette Williams completed the first successful well there, and other operators soon enjoyed similar triumphs.

Meanwhile, the underground gas discovered in Summerland in 1890 touched off a scramble for gas well leases. The Summerland Gas Company was quickly organized, drilling wells everywhere including in the town's streets on leases granted by Williams. Under public protest, he was forced to withdraw drilling rights from an out-of-town syndicate and the Gas Company. But the scuffle for gas leasing continued to disrupt the little community founded as a spiritualist center just a year before gas was discovered. Spiritualism was all but forgotten in the quest for natural gas.[12]

Oil wells, too, overwhelmed Summerland's earlier spiritualist inclinations during the first half of the 1890s. By 1896, there were twenty-eight to forty-two productive wells, and more were being drilled. Records were sloppy then, and production figures unreliable. The United States Geological Survey reported in 1906 that Summerland wells produced about seventeen barrels in 1895. Other reports put the number much higher.

No matter. The industry was on its way in Summerland. Most of the wells were inland, extending only to the coastal bluff. But in 1894 Williams drilled two wells on the beach, and added another in 1895. He discovered oil there, and it was inevitable the next step would be into the ocean. No one had ever drilled offshore, and oil workers had to invent a *new technology*. It was not pretty, but they did it.

Going Offshore

Whether the first wells were drilled from the Summerland piers in 1896 or 1897 is uncertain. Many details of this early chaos are in dispute because no one kept accurate records of an industry that was under virtually no public control or accountability. Drilled with the same rudimentary technology as used onshore, the offshore wells were the first in the nation, and very probably in the world. Twenty-two companies built fourteen piers before the end of the century, requiring the permission only of the oceanfront property owners and the county, and apparently county approval was automatic. Resting on wood pilings, the plank piers extended seaward as far as 1,230 feet. The pioneer companies drilled 412 wells from the piers by 1902, ranging from 80 to 400 feet deep.

"A typical Summerland well would begin with an output of five barrels each day but, within a year, it would be reduced by one-half and after two years to one-fourth of the initial output," according to David F. Myrick, a Santa Barbara historian. Just after the turn of the century, a storm battered the unstable wells, many of which were abandoned as relics. Only a few wells were still active in the late 1920s. The limited market was enough to absorb the total production of slightly more than three hundred thousand barrels from the wells on the piers.

The Railroads Helped

The Southern Pacific Railroad ran along the coast between the onshore wells and the piers, and the company became a producer, consumer, and carrier of offshore oil. The Southern Pacific Oil Company was organized as a producer, but it was not part of the railroad corporation. The railroad was involved in oil operations, however, through subsidiaries.[13]

Southern Pacific and Santa Fe railroads became consumers as they converted from coal to oil to run their locomotives. Gasoline was considered a waste byproduct until assembly lines began producing automobiles. In any case, the low quality oil was more suitable for paving, heating and lighting.

Because the railroad ran through the Summerland oil fields, much of the production was shipped by rail. Marine tankers, however, were making an appearance. The first tankers were not designed as petroleum carriers; they were modified Pacific whalers built originally to carry whale oil in large cargo tanks. Other vessels carried petroleum in barrels. The first ship designed as an oil tanker was Union Oil Company's *W. L. Hardison*. Launched in 1889, it went up in flames at the Ventura pier the same year. It was replaced by the tanker *George Loomis*, built in 1895 by the Pacific Coast Oil Company and the Union Oil Company.[14] But while marine transportation was available, it appears that most of the pier-produced oil was shipped by rail. Some of it did not go far. There were several small refineries along the Santa Barbara-Ventura coast.

Coastal Chaos

While production never reached exhilarating levels, the scramble for offshore opportunities was intense and sometimes violent. According to Myrick:

> The move of oil men to the beach and into the ocean offered additional space to work, but soon these areas were crowded by conflicting interests, as the beach operators and the wharf men sometimes occupied the same location near the shore.This was shortly after J. B. Treadwell found oil with a rig on one of the first piers.
>
> People were still talking about the great Treadwell oil well when the 'Summerland War' began and took up the first two weeks of August 1898. Reports are confusing and incomplete, with subtle innuendoes, but it does appear that Treadwell's wharf, in part situated on beach land claimed by others for their wells, was the target. T. D. Wood's platform was the first structure attacked and a portion was destroyed. Then Treadwell got into a fight with W. M. S. Moore, his other neighbor. Each party had armed guards, but some of these 'gunfighters' were handy with the axe. One of them chopped away a pair of supporting piles, which halted the pile driver's progress.
>
> The next day, the wood chopper hacked away at another bent [support] of piling, cutting any chances of retreat, so the pile driver was isolated. Fences and strings of barbed wire suddenly appeared to block access from the properties and one gate was secured by a lock with a closely guarded combination. Not even the guards were privy to this information, so entry and exit had to be made by shinnying up and down piling, which was unpleasant at high tide, to say the least. Some part of the Treadwell wharf was destroyed by the guard from other camps.[15]

Amid this turmoil, the waterfront was reduced to a blight. A California Coastal Commission staff report describes the waterfront at this time:

> [U]nrestricted private oil development led increasingly to wasteful drilling practices and pollution of the coastline, including the release of large quantities of natural gas into the atmosphere, the spillage and loss of oil due to poor drilling techniques, and the dilution of oil basins with groundwater—due to poorly conducted and uncoordinated drilling—rendering them unsuitable for further development.[16]

Oil storage was also sloppy.

> Along with other pumpers, Seaside [Oil Company] found they were pumping as much water as oil from their wells. After pumping they stored the oil and unwanted water in tanks near the wells. The water was allowed to drain out of the bottom of the tanks. This residue dripped onto the beach and into the ocean like an iridescent scum. In those bygone days, there were no laws that prohibited these types of spills.[17]

Even a passing glance (and whiff) from a train was enough to identify the waterfront as a place of dirty operations. Alfred Homan, editor of the *San Jose Mercury*, wrote about it in 1901.

> Summerland is a very different sort of place now, for in a day fatal to its peace somebody struck oil....[E]ven as the train whizzes by, one sees streams of black, syrupy looking oil pouring out from the pumps' mouths with the heavy flow and reluctant slowness of molasses running from a barrel on a cold morning. There are traces of Summerland's old character in the many pretty cottages about which the forest of 'rigs' has grown up, but the general transformation is painful to look upon.
>
> The whole face of the townsite is aslime with oil leakages....If my impracticable spirit were to have its way, the whole beautiful region bordering the Santa Barbara Channel would be reserved as a universal playground, never to be marred by the progress of industrial development.[18]

Santa Barbara editors agreed. Viewing the Summerland shoreline in 1899, the *Santa Barbara Daily Press* protested when a prospector wanted to explore for oil on the Santa Barbara shoreline.

> It would be an unfortunate disaster if the beach front near Santa Barbara's waterfront should be disfigured with the ugly derricks of oil wells. An attempt to force these unsightly creations upon the shore beyond Castle Point should be met by united resistance on the part of the people as a whole and the individual owners of adjoining property.

Other prospectors announced plans for seaside oil probing, but "met by united resistance," they gave up.[19]

That was in Santa Barbara. In Summerland, however, things were out of hand. Oceanfront property owners rushed to get into production before neighbors could suck oil from under their land. The state legislature took some action, but failed to stop the resource degradation wreaked by unregulated oil production. In 1915 the state Oil and Gas Division was created to encourage maximum production without wasting oil, but that did not get at the problem. Bolder steps were needed.

Regulation Fails

Without challenge from the federal government, the legislature in 1921 asserted that California coastal waters belonged to the state. Until then this offshore jurisdiction had been murky, but in the state Mineral Leasing Act of that year California claimed sovereign authority over the oil resources under the "marginal sea." According to the law the state surveyor general could grant offshore oil leases except in front of cities and one mile in either direction. But the law did not give the state authority to supervise offshore drilling. As amended in 1923 the act authorized the surveyor general to deny permits if oil exploration might damage nearby property. But even that authority was thrown out by the courts.

To test the constitutionality of the 1921 leasing act, the surveyor general denied some requests to drill exploratory wells in the Santa Barbara Channel. The California Supreme Court held that California owned the minerals under the coastal waters, but could not deny applications for leases under the terms of the 1921 act.[20]

The decision required the state to approve hundreds of ocean leasing applications while leaving the offshore oil industry unregulated. Between 1926 and 1929, more than two hundred persons applied for prospecting permits for parts of San Pedro Bay, San Pedro Channel, Santa Barbara Channel and Santa Monica Bay. The rush for requests increased after oil was discovered in the Ellwood area in 1928.[21] After about three hundred fifty unsupervised wells had been drilled under the 1921 act, legislators in 1929 said "enough!" Impelled by overproduction, coastal pollution, and visual blight, the legislature repealed the law and banned further leasing.

Debris Remains

Eventually the Summerland piers disappeared—or did they? Storms demolished some of them, some were removed by the owners. Many were left for time to take its toll. In 1997, a century later, the state was still cleaning up remnants of the Summerland offshore oil adventure. Santa Barbara County officials had found remnants of 165 wells that were drilled before 1907, and suspected that as many as 412 improperly plugged wells were underwater off Summerland, some of which were still seeping oil.

Timber from the Summerland piers rotted or floated away in time, but in the 1990s the state was still trying to remove long lengths of pipe, concrete and steel caissons six to eight feet in diameter, cement coated pipe, wellheads, steel wire, steel rails, H-beams, well casing, and other debris. Charlie Powell, a State Lands Commission engineer, noted:

> Some companies made an attempt to clean up by cutting wells at the sandline...but the sandline varies several feet as sand comes and goes with the storms and seasons. In an attempt to plug the abandoned wells, some companies put rocks, wood, or metal down the hole, but this didn't plug them. Some companies welded steel caps on the abandoned wells, but many wells were just filled with junk.[22]

So some leaked.

According to a 1994 county report, "the wells were blasted with dynamite and then plugged with telephone poles. This type of abandonment was not competent, and oil and gas eventually leaked through and around the casing....[T]he possibility of capping all the wells in the immediate future is remote."[23] (Chapter Two describes similar conditions off Ellwood west of Santa Barbara.) By the end of the twentieth century California taxpayers were expected to have been tapped for three to five million dollars for the cleanup.[24]

"In 1988 a high concentration of oil was observed off Summerland, and Lands Commission divers took a look," said Terry Bertoline, a commission supervisor. "They found at least three old wells that were still leaking in the surf zone."[25]

But the commission instead went after three leaking wells in the intertidal zone because they were exposed on land at low tide, at least during the winter when much of the beach was washed away. Surfers, who were most vulnerable to injury and pollution from the remnants, wanted the offshore debris attacked first. But the state did not know how to clean up leaking wells offshore. "We want to do onshore wells first where we can see what's going on; we're not quite sure how to go after the deep-water wells," said Jim Trout, assistant executive officer of the Lands Commission.[26] But even the leaking intertidal wells were too much.

In the spring of 1993 commission contractors erected an enormous steel frame platform and moved heavy duty industrial machinery over the bluff and onto the beach. Only two of the five intertidal wells were plugged before the $897,000 budget was exhausted. Taxpayers nationwide joined Californians in paying for the cleanup in 1995 when the Coast Guard indicated it would supervise the efforts.

Despite the chaotic conditions created and abandoned by the offshore oil pioneers, a monument to their achievements stands on a Summerland bluff. Dedicated in September 1978, it memorialized the "first offshore production on the Western Hemisphere...." The Petroleum Production Pioneers left room for a dubious Soviet claim of the first such production. For a time during the Cold War, the Soviet Union claimed credit for inventing almost everything.

The plaque was dedicated by Robert Lagomarsino, then congressman from that area, who said that "if we could let loose the spirit of those fields and get the government out of the oil business, many of our energy problems would be solved." Then he added: "With some restrictions, of course."

Chapter Two

Whose Ocean Is It?

Offshore oil activity at Summerland peaked and flopped quickly, but about the time it ended out there, oil was discovered elsewhere in the Santa Barbara Channel, this time west of Santa Barbara, and off the coast of Los Angeles and Orange counties.

With improved technology, primarily the rotary drill bit,[1] geologic formations under the ocean floor were probed by slant-drilling from onshore. With this technology oil was found in the late 1920s off Ellwood and Capitan in Santa Barbara County, and off the Rincon in Ventura County. The Wilmington field offshore Los Angeles County and the Huntington Beach field offshore Orange County were discovered from onshore wells in the 1930s. With one exception, no free-standing oil structures were built off the West Coast until after World War Two. The exception was "Steel Island" built by Indian Petroleum Corporation off Rincon in Ventura County. It was installed in 1932 and it went down in a storm eight years later.[2]

The biggest prewar discovery was at Ellwood, twelve miles west of Santa Barbara. Teaming up in 1928, the Barnsdall and Rio Grande oil companies drilled for four weeks on an oceanfront bluff until June 29 when they decided to abandon the field. The crew agreed to work through the daylight shift, and before quitting took one more core sample. This one had something new—Oil.[3]

To maximize their discovery, Barnsdall and Rio Grande planned one production well per ten acres. Producing at less than maximum at the outset would increase production over the long haul. But because the partners had lost some control through a faulty lease application, and because drillers on adjacent properties were less scrupulous and less professional, the competition for Ellwood oil became a scramble.

"The stage was set for the rape of Ellwood," said Charles S. Jones, then vice president of Rio Grande.

> The first offshore well at Ellwood was brought in on October 11, 1929, and within seven months 36 wells had been located on the beach or [on piers] in the ocean. Wells were produced wide-open.... More wells were drilled to increase production on one property, and to offset them others were drilled elsewhere. Some wells produced steadily at 12,000 to 14,000 barrels daily. Pressures soon dropped, and water flooded the rich sands.[4]

Jones said that because of wasteful production methods, one hundred million barrels were lost over time in the Ellwood field. But in 1929 Santa Barbara County's south coast was one of the leading crude oil producers in California. Hundreds of wells were producing along a seven-mile waterfront from Coal Oil Point to Naples, most of them on steel and wooden piers off Ellwood.[5] The producers were the largest taxpayers in Santa Barbara County. Charles S. Jones enjoyed telling about protesting the county's 1928 tax assessment on his company. The chairman of the Board of Supervisors is reported to have responded: "Charlie, I've got news for you. Santa Barbara County is going to build a new courthouse, and the Board of Supervisors have [sic] decided to give your company the honor of paying for it. Now please redo the tax bill in any way that pleases you, as long as you don't change the amount of tax on the bottom line."[6]

The offshore Ellwood field peaked in 1930 at 14,600,000 barrels of high-quality crude. Twenty-two more wells were drilled there the following year, but the field was already past its peak.[7]

As noted in Chapter One, the California surveyor general sought to gain some control over the chaotic oil activity by denying several hundred tideland prospecting permits between 1926 and 1929. On the last day of 1928, however, the California Supreme Court ruled that he was required to grant permits to qualified applicants. The legislature tried to correct this by prohibiting further offshore leasing.

That was not as prohibitive as it sounds. About 350 offshore wells had already been drilled, and they could stay active. Beyond that, drilling onshore at the water's edge, which was not banned by the state, drained oil from under the adjacent ocean floor. And by now drillers had learned to slant-drill, or whipstock, and could reach under the ocean from onshore wells. That was not legal, of course, but no one but the oil companies and their drilling contractors knew in which direction their drill bits had taken off. And sometimes *they* did not know. Wayward drilling from onshore in the 1930s tapped offshore deposits off Ellwood, Capitan, and Dos Pueblos west of Santa Barbara; Rincon in Ventura County; Wilmington in Los Angeles County, and Huntington Beach and Seal Beach in Orange County.[8]

Coastal Cities Resist

Appalled by the onshore oil drilling devastation in several oceanfront communities, other coastal cities hoped to fend off this assault. Redondo Beach, which owned the mineral rights off its shores, banned drilling there as well as onshore. The prohibition, enacted by a vote of the people in 1935, held for twenty years. Seal Beach, too, banned drilling onshore, and even claimed jurisdiction over its coastal waters, and forbade drilling there as well. Oil offshore Huntington Beach was being drained profusely and illegally by wells onshore that curved out under the state waters. Derricks lined the sandy beach between the Pacific Coast Highway and the water's edge. Upland from the highway, too, the city watched the oil rigs emerge like an ugly, instant forest.

In 1931, the legislature lifted the 1929 ban on new tidelands leases, and authorized offshore drilling permits. But the people overruled the legislature in a referendum, and then defeated initiative measures that would have approved new offshore drilling and onshore slant-drilling under the sea bottom.[9] Because oil tends to flow from untapped regions into nearby zones where wells are producing, onshore wells were draining oil from offshore. This was costly to the state but perfectly legal as long as the drill bit did not trespass under the tidelands. But many onshore wells *were* illegally slant-drilling under the ocean.

After three years of wrangling, the state and the oil industry in 1937 developed a royalty schedule for tidelands oil production. Part of the compromise was widely condemned as a state sellout. Standard Oil agreed to settle state claims for $500,000. State Senator Culbert O. Olson (soon to become governor), assailed the deal as "an outright gift by the state to the oil company of over $5,000,000 and a flagrant betrayal of the people."[10] In 1939, Attorney General Earl Warren (also a governor to-be) conceded at a congressional hearing that up to 1938 the state had not been diligent in conserving its oil. "It appears that this acknowledgment is a gross understatement of conditions, which. . .were chaotic," Ernest R. Bartley wrote in *The Tidelands Oil Controversy.*[11]

STATE LANDS ACT

After three years of unproductive state bargaining with the oil industry, the legislature in 1938 passed the State Lands Act and created the State Lands Commission to take charge of tidelands leasing and supervision. Its authority was limited. It could grant offshore leases only where oil was being drained away by onshore wells, or was under such a threat. Drilling was permitted from onshore wells, or from drilling islands. There could be no new piers and no new drilling from existing piers. Leases would be awarded in competitive bidding, and royalties would be on a sliding scale, depending on the rate of production. Armed with this authority, the commission approved drilling at Huntington Beach where more than a dozen onshore wells were draining the state's offshore resource. Now legitimized, scores of new drilling derricks soon penetrated the Huntington Beach shoreline to slant-drill under the ocean floor.

But by 1940, oil production from many wells on the beaches and piers had seen its best days, and the retreat was on. Again, the industry walked off and left a vast coastal junkyard—the remnants of piers, wells, and other industrial trash—for the state taxpayers to clean up.

The fifteen piers off Ellwood—some more than a half mile long—were structural improvements over the earlier ones at Summerland. And the state now required that leaking oil be contained where it was drilled and produced, and not slopped into the ocean. But effective public regulation still lagged years behind the pace of the industry's coastal enterprises. One truncated Ellwood pier, no longer supporting oil wells, became a landing for oil service boats. The rest were torn down, blown up, storm-damaged or left to rust and rot.

Incredibly, there was still no law or rule against trashing the ocean bottom. This was confirmed in 1966 by Frank Hortig, then executive officer of the State Lands Commission, who told a legislative subcommittee that abandoning oil industry debris on the ocean bottom violated no state law if it did not pollute the waters. Enforcing a ban on underwater debris would require a fleet of submarines, he said. Fishers, meanwhile, continued to rend their nets on seafloor junk.

Cleanup was left to the state. In 1979, the state hired Oceanographic Services Inc. to find the oil industry remnants offshore Ellwood, El Capitan, Goleta Beach, Coal Oil Point, and Isla Vista, all in the Santa Barbara Channel. The marine probers charted the remnants of piers and other industrial junk left after 1900-1920 at Summerland, and after 1920-1940 in five other coastal areas.

Most scrap was found off Ellwood, the most active oil field. A forest of old pier supports was visible at low tide after winter storms washed out the sandy beach. Unlike the wooden piers at Summerland, many Ellwood piers rested on steel supports that did not rot or float away. Even more astounding than the visible mess was the oceanbottom jungle of industrial relics farther offshore.

Oceanographic Services discovered more than 1,250 I-beams, and relics of wood and pipe pilings by the hundreds—perhaps more than one thousand. Fifteen steel caissons and at least a dozen wellheads were found, plus sheet metal, steel plates, rebar, cable and about seventy lengths of pipe. In a score of rubble piles trash was too entangled to sort out for identification. Junk at fifty-five other places could not be specifically described for lack of visibility.

According to the Oceanographic Services report, most piers were gone by the time of the survey, but

> many of the structures were cut off just below the waterline and the remaining portions protrude from the ocean bottom to various heights....These remnants in particular present hazards to boaters, swimmers, surfers and other users of the beach. Some of these hazardous obstacles have been removed since 1968 by the states and its lessees, though several remain.[12]

"Several" was an understatement. After the 1979 survey, the State Lands Commission made another of its sporadic attempts to remove the debris, but it was difficult and costly, and uncelebrated monuments to the offshore industry remained a century after it drilled the first offshore well.

Most of the tidelands prewar production came during the early 1930s, and made no major contribution to the World War Two effort. During the first half of the 1940s, strife among government, industry, consumers, workers, jobless, and others, subsided in a unified war effort. The big California oil story during the

Second World War was the February 1942 Japanese submarine attack on the Ellwood oil field. Though the attack did almost no damage it was a symbolic Japanese move which gains considerable drama with each retelling.

Of more relevance was the extended, convoluted post-war political war over who, in fact, controlled the coastal waters off California. Had California been leasing offshore oil fields over which it had no authority?

Whose Ocean Is It?

Although the courts had limited the state's authority to regulate offshore oil activity, no one questioned the assumption that the state owned the tidelands.[13] No one, that is, until Joseph Cunningham, who headed a California investment group, asked the Interior Department in 1934 for permits to drill exploratory wells. Cunningham argued that under the federal Mineral Leasing Act of 1920, Washington was the place to go for such permits. However, Secretary of the Interior Harold Ickes indicated that the state alone had tidelands leasing authority.[14]

Cunningham was adamant, and he induced Congress to look into tidelands ownership. Members of Congress debated the issue throughout the rest of the decade, but produced no legislation. During the debates, however, Ickes changed his mind. He now argued that states were more interested in enhancing oil industry profits and state revenue than in managing and nurturing natural resources.[15]

Congressional Debates

As the congressional debate wore on, the states lined up clearly on one side, and the Navy and Interior departments on the other. The Department of Justice, under Francis Biddle and later Tom Clark, appeared ambivalent; Ickes felt they lacked enthusiasm for going after the tidelands. The states, represented by their attorneys general, were organized, active and united. The oil industry backed the states which it believed to be more friendly to its interests. The Hearst press, at that time an intimidating political force in California, supported the state and the oil industry.

Ickes' resistance to state control was reinforced in 1944 when he believed that Edwin Pauley, a California oil producer, had attempted to influence Ickes and President Truman with a $300,000 oil industry contribution to the Democratic Party. In return Ickes and the president were asked to cease asserting federal title to the tidelands.[16] This episode did not surface publicly immediately, and the tidelands debate subsided during the first half of the 1940s as World War Two demanded full attention.

With the end of the war, the squabble resurfaced. Pressed by an insistent Ickes, the federal government in May 1945 sued to legalize its claim to the tidelands. The government demanded an injunction against Pacific Western Oil Company which was producing at Ellwood under a state permit. Ickes, however, wanted the federal government to sue the state of California, not a single oil company. He prevailed with President Truman, and in October 1945 Attorney General Clark dropped the suit against Pacific Western and took on the state of California.

Three weeks earlier, Truman had signed an order declaring federal control over the tidelands pending a decision in *U. S. vs. California* or future congressional action. He had already vetoed a congressional resolution for state control of the tidelands. But in January 1946 the president made a move that seemed to contradict this action. He submitted Pauley's nomination for Undersecretary of the Navy to the Senate. Ickes, fearing that Pauley would then soon succeed Secretary James Forrestal, did not want naval oil reserves under a cabinet officer who was actively engaged in the oil industry.

But Ickes kept his peace, and did not publicly oppose the nomination. When questioned by the Senate Naval Affairs Committee, however, he disclosed conversations with Pauley which he interpreted as attempts to influence offshore jurisdiction policy with offers of huge contributions to the Democratic Party.

Truman, of course, was incensed by Ickes' testimony, and although the nomination was withdrawn, he defended Pauley to the end. Under the circumstances Ickes could not remain in the cabinet. He resigned with a flourish in February.[17]

"Equal Footing"

Meanwhile, the federal government took its case for control of the tidelands to the Supreme Court. Defending its position, California insisted, *inter alia*, that the thirteen original colonies held rights to the tidelands under British rule, and that they retained these rights after independence. California, of course, was not one of the colonies, but argued that it was admitted to the union under the "equal footing" rule which conferred the same rights. The federal government replied that no rights to the coastal submerged lands had ever been conveyed to California or any other state. In 1947 the court ruled six to two (with one abstention) for the federal government, declaring that the thirteen original states never owned the coastal waters or the land under them.[18]

The ruling posed the problem of how to manage leasing, drilling and production that so far had been under state control and for which California had been collecting royalties. It was agreed that production should not be interrupted; the nation needed oil. The parties agreed that the state could continue to approve leasing and drilling with the permission of the Interior Department. The state would continue to collect royalties but hold the funds in escrow. In 1950 the state was required to deposit the royalties in a separate account in the United States Department of the Treasury. Although few if any new leases were awarded under this arrangement, drilling and production in existing leases increased and by 1951 more than 480 wells were producing from California tidelands.

How Far Out Is "Inland"?

While ruling for the federal government, the Supreme Court decided that the state still controlled "inland waters" along the coast, and left it up to a special master[19] to define the "outer limit of inland waters." California claimed the entire Santa Barbara Channel, extending fifty miles offshore to San Nicolas and San Clemente Islands, as "inland waters," and thus under state control. The special master, who took five years to present his recommendations to the Supreme Court, disagreed. He argued only bays along the coast qualified as inland waters.

But the tidelands dispute was now shifting politically in favor of the states. In 1952 Truman vetoed legislation that would give states control of the tidelands, but it was an election year and things would soon change. Dwight Eisenhower supported state control, and his election cleared the way for an entirely new kind of petroleum management off the nation's coasts. The "inland waters" dispute became moot—for the time being—when in May 1953 Congress passed the Submerged Lands Act which gave California authority over the ocean bottom extending three miles seaward. Congress not only returned control of coastal waters to the states, but also passed the Outer Continental Shelf Lands Act which established the rules for federal oil leasing beyond the three-mile line. In these waters, the Interior Department could offer leases for competitive bidding and collect bid money, rent for the leased areas, and royalties based on oil and gas production.

With its offshore authority restored, the state now had to bring order to its domain of "marginal waters." Free of effective control, the offshore oil industry for more than a half century had ravaged coastal resources. It polluted air, land and water, created a visual blight, and left dangerous and unsightly debris as each field was depleted and abandoned.

The state legislature debated leasing systems throughout 1954. In August it heard from determined residents at a Santa Barbara hearing conducted by State Senator James E. Cunningham.[20] With extraordinary unity and impressive presentations, Santa Barbarans demanded protection of their waterfront. With this backing, Assemblyman Stanley Tomlinson and Mayor John T. Rickard of Santa Barbara induced the legislature to exempt a thirteen-mile strip from Summerland to Goleta Point, where the University of California was building a new oceanfront campus.

In the Cunningham-Shell Act that was passed the next year, several other waterfronts also were spared from offshore oil leasing. Indeed, the act limited leasing to a stretch from the north end of Newport Beach in Orange County to six miles south of Oceano in San Luis Obispo County, minus some scenic sanctuaries such as Santa Barbara's. Even there, however, leasing would be permitted if the sanctuaries became threatened by drainage from adjacent federal leasing.

Elsewhere, however, the state could now offer leases even for areas that were *not* subject to drainage, an important change from the 1938 leasing act. The law also for the first time permitted freestanding offshore platforms and floating drilling rigs. Such rigs and ships were at work in the Gulf of Mexico, but nobody had yet seen anything like them off California.

The act also required the state to give notice to coastal communities where offshore leasing was planned, and the cities could demand public hearings. This procedure, routine today, was a sharp departure from current practice.

Reaching Out Again

With its renewed authority, California allowed its "inland waters" claim to lie dormant for ten years after Congress had returned the tidelands to state control. There was not any interest in—or technology for—producing oil from the deeper waters beyond three miles offshore. The Santa Barbara Channel bottom drops off to about two thousand feet deep in mid-channel. Technology improved rapidly, however, and by 1963 drilling rigs were ready to probe the bottom in deeper water. California was not about to let this new "land" slip away, and the state renewed its claim to the entire channel as "inland waters." Four of the channel islands are in Santa Barbara County, three in Los Angeles County, and one in Ventura County. Any waters between the mainland and the islands of these counties—and within the state—must be regarded as "inland," the state argued.[21]

Federal officials insisted that state control extended only three miles off the mainland and three miles around the islands, leaving much of the channel waters under federal authority since the islands range from eleven to fifty miles offshore. The feds won. In 1965, the Supreme Court denied the state's claim. Under the Submerged Lands Act of 1953, the court majority wrote, "the most plausible inference would be that Congress, in adopting the three-mile limitation, must have intended some base line to be used other than one dependent upon each State's subjective concept of its inland waters. . .."[22] The court upheld the findings of the special master in limiting "inland waters" to coastal bays. California's jurisdiction over oil drilling and production, then, remained limited to three miles offshore.

Since the 1965 ruling confirmed the existing arrangement, the state and federal offshore oil laws of 1954 and 1955 set the stage for a half century or more of industrial, political, and environmental revolution which would be centered largely on Southern California, but would be felt nationwide and around the world.

Chapter Three

Post-War Chaos

While state and federal agencies drafted new offshore leasing rules in the 1950s, the oil industry prepared to exploit new fields off Southern California. First, they had to locate and determine the borders of the fields which were farther offshore and under deeper water. This called for "advanced" technology.

The industry had found nearshore fields by drilling a few feet inland or seaward of the water's edge. To decide where to drill inland, seismic testing teams often exploded dynamite in shallow holes, then recorded the reverberations. This disclosed something about the composition, depth and shape of subsurface rock and soil. The system did not find oil—at best it found geologic formations that *might* contain oil.

It was not a perfect technique, but it was then the best available. So the industry took it offshore to investigate what lay beneath the ocean bottom. If it was not an ideal system onshore, it was a disaster offshore. An explosion in deep water formed a big bubble which collapsed instead of rising to the surface. This created a "second untimely train of pulses," said William Kennett, a petroleum geologist who recalled the experiments:

> To overcome this problem the charge was suspended [under water] from a balloon at a prescribed depth so the bubble would vent to the atmosphere. This created a spectacular water geyser and killed an unacceptable number of fish.[1]

Unacceptable indeed.

"Santa Barbara's birthright as a paradise for sport and commercial fishermen is being blasted to extinction," editorialized the *Santa Barbara News-Press*. That was May 2, 1948. The oil industry had been blasting "for weeks, in fact, for several years," the newspaper reported, and with the full cooperation of the State Fisheries Laboratory. Most coastal residents were enraged as the blasts came closer to shore.

According to Beth Porter, who lived on the ocean bluffs just east of Santa Barbara:

> We would often be awakened in the morning by the plates in our floor furnace rattling....I looked down from the bluff and I don't think you could lay a hand on that beach without encountering a dead fish. There was much protest, but instead of calling authorities, I called the editor of the Santa Barbara newspaper and the photographers were out there in minutes....That picture went all over the world....People were scavenging the fish for their freezers.[2]

It did not end with one dynamite run through an area. Because companies did not share their findings, each did its own blasting, subjecting some areas to repeated assaults. "Fishermen declare that from a few to a thousand fish are killed after each detonation, and that the blasts are occurring at frequent intervals day in and day out, over weeks and even years," the *News-Press* reported.

It's not that the state wasn't concerned. "The Fisheries Laboratory has a man aboard one of the boats whose duty it is to salvage the large fish that are killed [and rise to the surface]. He sells what he can and the money, ironically, goes into a fund for the preservation of fish.'"[3]

Fish were not the only victims; Santa Barbara homeowners said walls and foundations cracked under the percussion. Some city council members were aware of the structural damage, but complained only of fish loss when they wrote to Governor Earl Warren. Attached were petitions signed by more than sixteen hundred irate citizens.

The city got the attention of the California Fish and Game Commission which held hearings, and the State Lands Commission which indicated that from now on oil companies must pool their seismic tests instead of each doing its own. Blasting was banned for a time, then resumed under tighter controls. The oil companies also agreed not to blast offshore the city of Santa Barbara, within a mile of the shore from the Ventura line to Gaviota, nor within a quarter-mile of shore west of Gaviota.[4]

The blasting ended, eventually, when the industry found better ways to test the subsurface of the ocean floor. By the mid-1950s, seismic testing and ocean-bottom coring had reached a greater professional level (and depth). By one seismic technique or another, the industry confirmed it was floating over 1,800-square-mile mine of black gold.

Floating Drillers

But without leases, oil companies could only investigate the ocean floor through seismic testing or core sampling. They could not drill full-bore exploratory (wildcat) wells which could lead to production. Under the 1938 state leasing act they could drill under the tidelands only from manufactured islands or by slant-drilling from onshore.

The first floating core-drilling rigs were unwieldy, but creative engineers condensed a couple generations of progress into a few years. After World War Two the Navy made available surplus mine sweepers and patrol boats which the oil industry purchased and converted to drilling craft. The first vessels drilled only small and shallow core holes in the quest for traces of oil.

In 1950, W. W. Rand, an engineer and geologist, designed and installed a drilling rig mounted over the side of a 173-foot ex-Navy patrol boat. He named the ship and his company Submarex, and produced several more drillships. But unstable over-the-side rigs were soon abandoned in favor of drilling through a hole in the middle of the vessel. With this greater stability, wells were bored three thousand to four thousand feet beneath the seafloor.

"By 1954 the art of drilling for geological information in the seafloor had developed to the point of being practical," said William Kennett, "but as the time was approaching when areas would be opened for leasing through competitive bidding it became desirable to acquire additional information at greater depth and in deeper water."[5]

In 1955, four oil companies pooled their funds and engineering genius to convert a bigger vessel to drill in deeper water. They bought a 260-foot ex-Navy ship, removed the engines, carved a "moon hole" through the middle, and installed a 98-foot derrick over the hole. Lacking self-propulsion, it was towed into position by tugs. The driller was named *CUSS I* to represent its four owners: Continental, Union, Shell, and Superior oil companies. Operating in water depths of 50 to 350 feet, it drilled as deep as 6,000 feet under the ocean floor.

CUSS I did not include Standard Oil of California, but the company was by no means out of the picture. In 1955, it bought and converted a 204-foot ex-Navy landing craft. That became the *Western Explorer*, and right behind that came another conversion job, the *Jessie Andrews*, by the Richfield Oil Company.

"A peak in the exploration program was reached in 1956-59," the Geological Survey reported. "New methods and improved equipment made it feasible to drill to depths of 7,500 feet and to operate in waters as much as 400 feet deep."[6]

Discoveries and Production

The modern offshore oil industry in California was spawned by a 1957 amendment to the 1955 state leasing act that permitted drilling wells from offshore vessels, and producing oil from free-standing platforms offshore. The companies scrambled for permits, and in the late 1950s and 1960s more than thirty parcels were leased, mostly in the Santa Barbara Channel extending nearly seventy miles from West Montalvo field in Ventura County to Point Conception in the southwest corner of Santa Barbara County. By 1969, all the tidelands in the channel except the sanctuary off Santa Barbara was under oil industry contract, and twenty-four oil and gas fields were in production.

Most Santa Barbara Channel fields were discovered by drilling from the new vessels. But many fields south of Los Angeles were discovered in quick succession by slant-drilling from onshore. These included fields off Venice, Torrance, Long Beach (Wilmington field), Seal Beach, Huntington Beach, Newport Beach, and Redondo Beach.

Over the protest of the city of Seal Beach, which claimed jurisdiction in the adjacent tidelands, the Monterey and Texas oil companies wanted to build Monterey Island a mile and a half offshore in the Belmont field. Construction started after the California Supreme Court ruled that the city did not have the jurisdiction it claimed. The island was built of interlocking sheet steel pilings forming a wall 75 feet in diameter that was filled in with rock, sand, and wood piles. It began producing in 1954, was modified in 1955 to boost production, and by 1960 had 41 wells going. Island Esther, with spaces for 128 wells, was built by Standard Oil of California a mile and a half off Seal Beach in 1964-1965, also in the Belmont field. A storm in 1983 destroyed it, and Platform Esther was built atop the rubble.

Monterey was the first man-made, rock-filled drilling island in the world, and was followed shortly by another built by Richfield Oil Corporation in 1957-1958. That one is still a conspicuous landmark off Highway 101 at Rincon Point between Santa Barbara and Ventura. It covers 2.5 acres at water level, and stands 41 feet above the ocean. With its palm trees, half-mile causeway linking it to shore, and marine biology lab, Rincon Island undoubtedly will stand long after production ends.

Something else was taking place off Rincon Point that was not so obvious; it happened under water. Richfield installed California's first oceanbottom wells, or subsea completions, resting out of sight on the ocean floor. They were drilled from vessels on the surface, but when drilling was completed, production machinery was installed under water and was controlled remotely. This eliminated the cost of a platform, and it did not blight the seascape. But the system was less convenient for reentry, redrilling, or new drilling.

Richfield installed similar wells on the ocean bottom in two lease areas off Ellwood in 1961. Seafloor wells were also placed nearshore the next year in the Alegria and Gaviota fields west of Santa Barbara. These fields were not big producers, and some wells generated only natural gas. Seabottom wells were connected to processing plants onshore by oil and/or gas pipelines along the ocean floor.

Then Came the Platforms

Seafloor wellheads were out of sight, but the stark shape of things to come was dramatized on June 15, 1958 when Standard Oil of California towed Platform Hazel to a spot two miles off Summerland-Carpinteria.[7] The 160-foot steel framework was towed in from San Diego without benefit of a barge. The steel framework floated because the bottom of the legs were airtight caissons 27 feet in diameter. Indeed, it floated so high that ballast was required to keep it stable. After a few glitches, it was nudged into position, and the leg tanks were flooded. They gravitated to the bottom in one hundred feet of water, and were driven more than twenty feet into the ocean floor.

The twelve thousand square-foot working deck sat one hundred feet above the ocean. A barge as long as a football field with an enormous crane lifted the superstructure onto the steel frame. When a 160-foot drilling tower was added, the structure rose more than twenty stories above the water. It was two miles offshore, but with an unobstructed view across the water, two miles appears to be less than half that distance. It was a conspicuous addition to the seascape previously interrupted only by low-profile vessels and the distant Santa Barbara Channel Islands.[8]

Hazel was not alone for long. Platform Hilda was placed two miles west of Hazel in August 1960. Like Hazel, it had slots for more than twenty wells, and was in about one hundred feet of water about two miles off the Summerland-Carpinteria waterfront. Both were operated by Standard Oil of California, with Humble Oil and Refining Company as a financial partner. (The companies were later reorganized as Chevron USA and Exxon USA, respectively. They were, and are, part of the Standard Oil family.)

As oil flowed freely and profitably through more than forty wells, the two companies explored other parts of their leases in the same area. In 1966, they installed Platforms Hope and Heidi, each with slots for sixty wells, right up against federal waters three miles offshore.

Other companies, meanwhile, installed platforms in the western end of the Santa Barbara Channel. Texaco set Helen in ninety-five feet of water west of Gaviota in 1960, and three years later placed Herman east of Point Conception. Phillips Petroleum converted a drilling vessel into a production platform, named it Harry, and installed it near Point Conception in 1961. These three were modest producers that peaked early, pooped out rapidly, and were removed in the 1970s and 1980s.

One more company found the state leases worth a platform in the Santa Barbara Channel. In 1966, the Atlantic Richfield Company which had just evolved from a merger of the Richfield Oil Corporation and the Atlantic Refining Company, placed Platform Holly two and one-half miles seaward of the old Ellwood piers site. ARCO operated the platform, but Mobil Oil held a fifty percent financial interest.

Holly was the eighth and last platform set in state waters in the channel, but the industry also installed wells off Orange and Los

Angeles counties. The Huntington Beach offshore field had long been tapped by onshore wells, legally and otherwise, and drilling derricks lined the broad sandy beach between the Pacific Coast Highway and the water's edge. In 1960, Huntington Beach had a population of 11,000. The oil industry had room to grow without arousing stout opposition. In 1963 Signal Oil extended its reach seaward with Platform Emmy a mile and a quarter offshore. It had slots for forty-nine wells.

The next year Union Oil of California, having had no luck in the Santa Barbara Channel, built Platform Eva in Houston, towed it on two barges through the Panama Canal and tipped it off in fifty-eight feet of water a mile west of Emmy.

The big oil event of the 1960s, however, was centered less than a mile offshore in the Long Beach Harbor. Everyone knew that the Wilmington oil field, the state's largest, extended under that water; no seismic survey was required to find it. But there were obstacles to producing from offshore. Long Beach, hemmed in on three sides by forests of onshore derricks, was not in love with the encroaching oil wells, and had banned drilling in the city and in the harbor. The concern was well-founded. Oil and water were being extracted at such volumes that the city was sinking. The federal government, which had its own waterfront installations, sued Long Beach and the state to do something about serious subsidence. In 1959, Long Beach built huge water-injection plants that satisfied the federal government. In 1965 the oil industry persuaded Long Beach and the state, which shared jurisdiction over the area, to offer oil leases in the harbor.

Eighty percent of the new lease area was put up for bids in the first sale. The lease was won by a consortium of five companies collectively called THUMS, representing Texaco, Humble, Union, Mobil and Shell.[9]

Because the water was only twenty-five to forty feet deep, the THUMS partners built four ten-acre islands instead of installing platforms. The islands are low profile, but oil-drilling derricks are high and conspicuous. To soften the industrial appearance, walls were built around the derricks to resemble high-rise office or apartment buildings, enhanced with waterfalls and landscaping. Lighted at night, they still stand as tourist attractions. Production from several hundred THUMS wells peaked in 1969 at nearly sixty-four million barrels.

Onshore Industrialization

Once extracted, tidelands oil came onshore for processing. And every producer had its own plant. Along the Santa Barbara Channel each company picked out a canyon and industrialized it to process crude oil, natural gas, or both. Along the Santa Barbara coast alone, thirteen industrial plants were built at ten sites to handle offshore oil. There were no refineries; only processing plants that removed water, natural gas, and condensates. Gasoline and other crude-based products came from refineries in the Los Angeles, Kern County, and the San Francisco Bay areas and one small refinery in San Luis Obispo County. Offshore producers also needed storage tanks, and many had moorings for tankers that took on oil through pipelines from onshore tanks. Some companies had pipeline connections to Los Angeles area refineries.

The waterfront plants were built under considerable public protest, and with little public control. Regulations requiring clean operations and responsible abandonments were absent, weak, or unenforced. Many plants soaked the soil deep with oil and toxic wastes. In 1986, for instance, PCBs, petroleum and other pollutants were found several feet down into the soil at the Shell Western gas processing site twenty miles west of Santa Barbara.

By the mid-1990s, some ancient coastal oil plants had been abandoned and removed, and others were living out their last years. In the meantime, the federal government sabotaged any hope of bringing order out of chaos when it followed the state's example and began leasing federal waters off Southern California.

The feds dive in

The first federal foray did not amount to much at the time. Some companies, especially Shell, had been probing beneath the ocean bottom off Central and Northern California, and were eager to drill exploratory wells in 1963 when the Interior Department offered leases. From Point Conception to the Oregon border,[10] fifty-seven leases covering more than three hundred thousand acres were sold for bids totalling $12.8 million. The companies drilled twenty exploratory wells, and nearly every one turned up oil. But the price for the poor quality oil was too low to cover production costs. The crude from the onshore Santa Maria Basin in Santa Barbara County was selling for one dollar a barrel. Union Oil's onshore field could make a profit at that price, but it would not support a costly offshore enterprise. The industry abandoned production plans. So fifty-seven offshore leases came and went quickly with hardly a murmur of public concern, pro or con. But anger would soon replace apathy.

Although the Santa Barbara Channel was now known to be rich in oil, the federal government had not offered leases there before 1965 because its authority was challenged by the state of California. When the Supreme Court ruled that the state's authority extended no farther than three miles offshore, the Interior Department was free to get into the oil business there.

If it had been left strictly up to Secretary of the Interior Stewart Udall, there probably would have been little oil activity in the Outer Continental Shelf off California. In the foreword to his book, *The Quiet Crisis*, he wrote in 1963:

> America today stands poised on a pinnacle of wealth and power, yet we live in a land of vanishing beauty, of increasing ugliness, of shrinking open space, and of an over-all environment that is diminished daily by pollution and noise and blight.
>
> This, in brief, is the quiet conservation crisis of the 1960s.
>
> It is not too late to repair some of the mistakes of the past, and to make America a green and pleasant—and productive—land. We can do it if we understand the history of our husbandry, and develop fresh insight concerning the men and the forces that have shaped our land attitudes and determined the pattern of land use in the United States.

But Udall faced pressure from several sources. He was surrounded by a staff whom he trusted, but who had exploitive notions about the Outer Continental Shelf. After the Supreme Court ruled in May 1965, the federal government lost little time going after the OCS oil. William Pecora was appointed director of the United States Geological Survey on September 27, and two weeks later he proposed a coring program to probe the geological formations under channel waters. Udall approved the program on November 3. The federal government was impelled by at least two incentives.

Standard Oil of California was pumping from the outer edge of its state lease in the rich Carpinteria offshore field. Because the field extended beyond three miles seaward, Standard's production from two platforms was obviously draining oil from federal territory. Consequently, the Interior Department wanted a platform on the federal side to collect its share of revenue in bids, lease rent, and royalties from the field. That was the federal foot in the door, and soon the White House became interested in the entire Santa Barbara Channel, largely because it needed revenue to pay for the costly military intervention in Vietnam.

Luther Hoffman, representing Secretary Udall, came to Santa Barbara to advise local officials that the federal government had leasing in mind because it was losing oil to production in state waters.

Mutual Fears

As Sacramento and Washington each feared losing oil to the other through drainage, the implications were obvious. Washington was already impelled to action by this fear, and the state in turn probably would open the sanctuary off Santa Barbara for oil drilling if the federal government authorized production nearby. In January 1966 the city of Santa Barbara called on Congress to protect the state sanctuary by creating a federal buffer zone immediately seaward. Indeed, after the 1965 Supreme Court ruling, the City Council and the County Board of Supervisors petitioned Congress to make all federal waters of the channel an oil-free sanctuary.

The protest at that time was limited largely to activity in the Santa Barbara Channel, and based generally on the threat of visual blight and the oil spill potential. Buy many in the land of good living also recognized a far greater ecological threat. The natural resources were extraordinary—and vulnerable—along the sweeping two hundred fifty-mile California coastal arc embracing a profusion of marine life from Point Conception to the Mexican border.

Many marine species of the northern and southern Pacific Ocean habitats meet, and overlap for some distance, at Point Conception. That, together with a lavish supply of nutrients thrust up by marine upwellings, attract an unusual number of ocean creatures and a rich diversity of species. Commercial and sports fishing are major industries.

Many nearshore marine animals find food and shelter in the giant kelp forests which also are valuable resources for food and chemical industries. Kelp is harvested down to a maximum permissible depth below the water surface. This can be done often because kelp grows rapidly, often to lengths of two hundred feet. Southern California kelp beds shelter twenty to thirty fish species that are common in that habitat, and one hundred more are there seasonally or occasionally.

Despite this attraction, most California coastal fish live beyond the kelp beds. About 560 species thrive in California coastal waters. That is an impressive food and commercial factor, but emotions are more aroused by the big sea creatures—whales, porpoises, dolphins, seals, and sea lions. These are not fish; they are marine mammals, no longer consumed by Americans except in their hearts.

Twenty-five or more species of cetaceans—whales, dolphins and porpoises—inhabit these waters. From ship and shore, recreational whale-watching became big business. Californians watch as several kinds of creatures, now protected, struggle for survival of their species. The blue whale, the largest creature ever to inhabit the earth, is in danger of extinction. Its dimensions—up to two hundred feet long—are appreciated only when one occasionally dies and washes up on a beach.

Another group, the pinnipeds, attract similar attention and support. These are seals and sea lions, also once nearly hunted to extinction, now fighting their way back. Some, as the California sea lion, proliferate. Others, as the Guadalupe fur seal and the Steller sea lion, are still struggling. San Miguel Island in the Santa Barbara Channel has a pinniped rookery enjoyed by six species of seals and sea lions.

The sea otter, smaller than cetaceans and pinnipeds, is the darling of many coastal Californians because of its entertaining diving, foraging, and grooming antics in the kelp beds. Abalone divers, however, despise the creatures as voracious competitors for valuable shellfish.

And there are birds; more than 450 species on the beaches, in the rocks and cliffs, on the water, on the islands. All eight Santa Barbara Channel Islands are significant bird rookeries. Birds by the millions live on the continent's edge, but some of them, too, were listed as endangered species. That included the imposing California brown pelican whose primary rookery north of Mexico is Anacapa Island in the Santa Barbara Channel. Eight hundred to one thousand chicks are hatched there annually, but near-disaster awaited. In 1970, a single chick survived eggshell thinning brought about by DDT in the adults' bodies.

Coastal and marine wildlife habitats range from sandy beaches and sandy bottoms, to rocky coasts and hard bottoms; from coastal cliffs and caves to estuaries, wetlands, and intertidal washes. All are vital spawning, foraging, and migratory regions. Many have recreational, scientific, commercial, cultural, archaeological, artistic, and spiritual values. Because of their vulnerability, some were protected as sanctuaries. But in the 1960s as Big Oil approached from seaward, Santa Barbara wanted more sanctuary protection, and sought help in Washington.

But almost nobody in Washington was listening. In May 1966 the Department of the Interior's Bureau of Land Management recommended a lease sale to prevent further drainage loss to Standard Oil off Carpinteria. In September Interior Secretary Udall said that one nine-square-mile lease tract would be offered to the highest qualified bidder. In December a coalition of Phillips, Continental, and Cities Service oil companies was awarded the first federal oil lease in the Santa Barbara Channel for slightly more than $21 million. The lease was effective New Year's Day 1967.

Year of Negotiations

Santa Barbarans strongly suspected that this was just the beginning. It opened a year of frantic negotiations between powerless city and county officials, and Interior Department administrators who felt no obligation to coastal communities. Within the Interior Department, Udall was the most sympathetic with seashore dwellers. But as he explained later, he relied on his staff which assured him that oil could be produced in the channel without degradation. He conveyed this assurance to Santa Barbarans.

Concern about oil at sea, already high, was intensified in March when the *Torrey Canyon* supertanker, carrying about 120,000 tons of Kuwaiti crude, went aground on a reef off the Scilly Isles. Most of its eighteen cargo tanks were ripped open. The news was full of oil pollution on the high seas.

Local delegations flew to Washington. Federal officials flew to Santa Barbara. Santa Barbarans asked the federal officials to forget about leasing, or at least delay it until a federal sanctuary was in place seaward of the state sanctuary, preferably all the way across to the Channel Islands. The California Senate asked Congress to require federal officials to at least confer with local authorities before offering more leases.

In September, a three-member Interior Department team[11] flew to Santa Barbara and offered to create a two-mile wide ecological preserve seaward of the state sanctuary. The Interior team wanted a quick response to the offer because the department was ready to submit a half million acres in the channel for bidding on October 15, hardly three weeks off. This would be the first major federal oil lease sale in the channel. The county wanted leasing delayed for six months while it contemplated the overwhelming implications.

With no more time for another cross-continental junket, the Board of Supervisors phoned Secretary Udall who agreed to a sixty-day postponement.

The county used that time to draft a report urging Udall to scale back the lease sale drastically since it would be an experimental venture, and the county and federal governments had better know what they were doing before they plunged into a monumental offshore industrial enterprise. Many questions remained unanswered, and nervous Santa Barbarans felt that much could be done to reduce the potential damage if the Interior Department were not in such a hurry to immediately open a vast area.

No law required the Interior Department to listen to local governments, and the Santa Barbara County supervisors' last trip to Washington in November 1967 was, in their minds, a waste. Some Ventura County officials and business leaders felt otherwise. They followed the Santa Barbara group to Washington to urge the Interior Department to go to bids as soon as possible.[12] Ventura County relied more on the oil industry, with many production and oil-service enterprises and its deepwater industrial harbor at Port Hueneme. Leasing was proposed off Santa Barbara, not Ventura, but Ventura stood to gain economically. The Santa Barbara County group was devastated.

George Clyde, at the time a member of the Board of Supervisors and a leader in the effort to save the channel from petroleum industrialization, commented some time later:

> None of our major requests was granted, except the original buffer zone....We asked for an extension of the buffer zone eastward to further protect the [state] sanctuary which, incidentally, if it had been granted, would have prevented the drilling of the blowout well....It was obvious to all of us involved in the long discussions during 1967 that everyone in the mineral resources section of the Interior Department was hell-bent to lease this channel and get drilling started.[13]

Outnumbered

It was not only the Interior Department. The shape of things to come was unveiled blatantly even before the county delegation left for Washington in November. Yielding to public protest, the Army Corps of Engineers agreed to conduct a hearing on whether Phillips Petroleum could install a platform in the Carpinteria Offshore field. The *News-Press* report of the November 20 hearing noted proponents outnumbered opponents about four-to-one. State and federal government agencies and the oil industry turned out impressively, bringing in witnesses from Sacramento, New Orleans, Los Angeles, and Washington. The opponents were home-grown, representing the Sierra Club, League of Women Voters, improvement and protective associations from Montecito and Summerland, the mayor of Carpinteria, and others.

State Fish and Game Department representatives maintained the platform would be good for fishing, the Coast Guard saw no problem, the Army Corps of Engineers was impatient with the opposition, and the Geological Survey was more than eager. Donald Solanas, regional supervisor for the USGS, told the hearing panel:

> It is of critical importance that there be no delay in the installation of this platform. The necessary platform approval has been obtained by my office which is responsible for and supervises oil and gas operations on federal lands.[14]

Intensive public concern assured that the hearing would be heated, and it was. But it was nothing compared to the public response when the audience learned that the platform was already being towed to the Santa Barbara Channel. The hearing was exposed as a sham. The battle over Phillips Petroleum's Platform Hogan was over, and the opponents had lost.

But now the BIG lease sale loomed, set for February 6, 1968. Secretary Udall came to Santa Barbara to report that the leasing would be conducted under the highest safety standards.

Big Time Bidding

By this time, Big Oil was sure there were rich deposits in federal waters off Santa Barbara, and the lease sale was a major attraction. Bids were opened at the Biltmore Hotel in Los Angeles. Offers were submitted for seventy-five tracts, seventy-one of which were leased. All the bids submitted amounted to nearly $1.3 billion. Total of the high bids was $603 million, a world record.

Robert Easton, in his book *Black Tide*, observed:

> Oilmen were smiling....so were Department of Interior officials sitting at the head table with Grant. [William E. Grant, head of the West Coast office of the Bureau of Land Management, conducted the lease sale.] Among them was Eugene W. Standley, who has attended all top-level meetings on channel oil matters and who had probably done more than anyone else to see the leasing through to this remarkable conclusion.[15]

But it did not impress the locals.

"The dollar amount of the bids doesn't mean much to average Santa Barbarans, most of whom are disappointed that any leases will be granted," wrote Walter Healy of the *Santa Barbara News-Press*. "The concern locally is the location of the leases."[16] Drilling rigs, he wrote, were expected "in the next few days."

And so they came, a fleet of bizarre vessels, their tall derricks oscillating in the swells. *News-Press* management was impressed, albeit negatively, and decided that offshore oil required close journalistic attention. I got the assignment, not because I knew anything about it, but because I had just been yanked off another assignment and had suddenly become available. *News-Press* photographer Wally Stein took to the air for shots of the oceangoing drilling rigs, and next day dumped a stack of prints on my desk. The city editor asked me to write captions for the pictures. "I'd be delighted," I said. "What the hell are they?"

I learned fast.

These were a new generation of drilling vessels, no longer converted Second World War surplus naval ships. They were designed from scratch for deepwater oil well drilling. Three very different kinds of floating rigs turned up:

• Drillships. These were self-propelled vessels with a derrick amidships over a vertical opening in the hull, known in the trade as a moonpool. Capable of moving under their own power, they were quite mobile, but they lacked stability in rough seas. They did well in protected waters such as the Santa Barbara Channel. During drilling, each ship was held in place by a ring of anchors and/or by dynamic positioning in which forward, reverse and side thrusters (propellers) were activated by computers to keep the vessel in position.

• Semi-submersibles. These vessels were towed into position, although later models, not used in the channel, had their own propulsion source. When in place over the drill site, hugh vertical legs and lower hulls were flooded, lowering the drilling deck to near the water level, and providing tons of ballast for substantial stability. They, too, could be held in place with anchors or dynamic positioning. To move the rigs, water was pumped out of the tanks and the vessel rose and floated on the lower horizontal hulls and could be towed.

• Jack-up Rigs. These strange creatures had three or more long steel framework legs that could be jacked up and down through wells in the hull or outside the hull. When towed with the legs in the up position, the vessel was topheavy. When the rig was in place, the legs were jacked down until they reach the ocean bottom, and the hull—or working deck—rested on the legs rather than floating on the ocean surface. They could drill only in water shallow enough for the legs to reach bottom.

These vessels were drilling rigs, not platforms. Where drilling discovered enough oil, permanent production platforms were brought in and secured to the ocean bottom. Platforms were usually much larger than drilling vessels. Wells were also drilled from platforms, but they were usually for production, not exploration. Scores of wells could be drilled from each platform.

The press seldom distinguished between offshore exploratory drilling vessels and production platforms. Everything was a "drilling rig" to most news reporters, few of whom specialized in offshore oil activity. News items were often written by reporters unfamiliar with the terminology, as I was in February 1968.

But the industry and federal officials invited me to visit drilling rigs, fly over the channel, and chat over coffee to familiarize me with fast-moving events. As many as six or eight drilling rigs were in action at any time in the channel, but the demand was always greater than the supply. The vessels were owned by drilling companies, and oil producers vied through competitive bidding to charter the rigs.

A Year of Spills

During the year of intensive drilling there were enough oil spills in the channel and elsewhere off the Southern California coast to keep the *Santa Barbara News-Press* and its readers edgy and contentious. Union Oil set the stage when it was fined for dumping oily waste into the Los Angeles harbor in November 1967.[17]

Then just as the hotly-disputed Platform Hogan was about to begin production in June 1968, it spilled more than two thousand gallons of crude into the ocean less than four miles off Carpinteria. The *News-Press* heard about it from a swimmer and editorialized:

> [T]here appeared to be no inclination on the part of the oil company [Phillips Petroleum] or state or federal officials to advise the public of the fact of the accident or the nature of the investigation going on....So far as the public's right to know fully and immediately about oil drilling accidents in the Santa Barbara Channel, a poor precedent has been set in this case.

A greater precedent than anyone knew at the time.

Leakage and accidents in the offshore oil industry disturbed Assemblyman Winfield Shoemaker, chairman of the State Assembly Subcommittee on Marine Resources. Representing Santa Barbara County, he scheduled a hearing on what he said were sloppy operations. "Recently there have been several serious incidents involving the oil industry and damage to other marine resources," he said, listing a few:

> Only last week [early July 1968] some 3,000 barrels of gasoline leaked into the ocean at Gaviota, doing serious damage to marine life, particularly lobsters.
>
> Last month, 2,000 barrels of crude oil went into the ocean off Carpinteria. [Previously noted.]

> In January, a tanker at Morro Bay pumped 2,800 barrels of oil into the ocean, seriously damaging the marine ecology of the area.
>
> In June, a pleasure boat hit an unlighted oil company marker buoy at night and sank. Fortunately, the two passengers were rescued.

For some reason he did not mention a tanker-freighter collision in the Santa Barbara Channel less than a month earlier. The 523-foot military tanker *Cossatot*, carrying 127,000 barrels of jet fuel, collided nearly head-on with the *Copper State*, a 492-foot freighter, twenty-five miles off Santa Barbara and about ten miles from Union Oil's Platform A. Taking the blow in the bow, the *Cossatot* was spared any rupture of its cargo tanks.

None of these events was major, but the series of mishaps put the community on edge. They demonstrated the potential for disaster. Anti-oil protests mounted throughout 1968. The Sierra Club condemned plans for oil drilling off San Miguel Island while Congress was being asked to create a Channel Islands National Park. The park did not materialize until 1980, but the dream was there for a long time.

County, state and federal agencies ruminated over who would be responsible in the event of spills. No one was sure who had authority under varying circumstances, and Congressman Charles Teague, representing the channel counties, introduced legislation aimed at clarifying these matters. The county grand jury foreman called for greater safety requirements after two oil industry helicopter crashes that took nine lives, and after some workers were killed in falls from boats and drilling rigs.

Onshore Standards

Meanwhile, the county Planning Commission and Board of Supervisors worked on standards for onshore oil installations, including landscape design, noise abatement, grading rules, industrial siting criteria and other regulations to protect recreational and scenic resources and residential quality.

Just before Assemblyman Shoemaker's legislative hearing opened in September, the Western Oil and Gas Association (WOGA) announced it had no representatives available to testify. They had "more pressing" meetings to attend, said Henry Wright, WOGA's executive secretary. Any demand for greater controls irritated the industry, and on one occasion Wright took it out on the county Planning Commission.

"A lot of beautiful raw land is worthless unless the people can use it, buy it, build on it and provide a place to work on it," Wright said.[18] A *News-Press* editorial interpreted his comment this way:

> What he was saying, of course, was that some part of the oil industry is annoyed because Santa Barbarans have tried to direct, limit and guide the activities and facilities of this industry in such a way as to minimize the damage to the scenic, esthetic, recreational, residential and tourist-attracting values of the channel, the tidelands, the beaches and the shoreline. If this is resented by some oil men, so be it....[19]

If the oil industry found county controls too stringent, many Santa Barbarans thought the Planning Commission and the Board of Supervisors were giving away the store. The county had rezoned land near Carpinteria for a Humble Oil processing plant. A citizens committee, headed by former State Senator Alvin Weingand, collected enough signatures to put the issue on the November 1968 ballot. The voters of the city of Carpinteria favored the plant, as did a majority in northern Santa Barbara County. But the county's south coast (outside of Carpinteria) opposed it and when the votes were counted the processing plant was doomed, 44,290 to 41,404.

Offshore, however, events were moving rapidly as the drilling flotilla discovered rich deposits one day and dry holes the next. In March, Union announced a big strike on lease tract 402, the area the county had said should have been in the federal buffer zone next to the state sanctuary. Union was the operator for a coalition of four companies that had bid $61.4 million for the three-mile square lease 5.5 miles off Santa Barbara. The other partners were Mobil, Texaco and Gulf.

Fatal Waiver

Two months later Union asked the Corps of Engineers for approval to erect two platforms in the lease area. No environmental review was required. The Corps was concerned primarily with navigational safety, and it did not even have to conduct hearings. It could make the decision without any help from the citizenry. Fred Eissler of the Sierra Club launched a drive to require hearings on requests for any new platforms, and several other groups joined the campaign. But in July the Corps of Engineers indicated it found no reason for more hearings. It had already held a hearing, according to Jacob S. Greenfield, counsel for the Corps' Los Angeles office. The November 1967 hearing on Platform Hogan, he said, covered the public's concerns.

To the Corps of Engineers and the Interior Department, public hearings on channel oil operations were a damned nuisance, to be avoided in the most effective manner. For expert advice, officers of the Corps consulted Eugene Standley who had done more than probably anyone in the Interior Department to move the leasing program with the least possible public interference.

> I pointed out that we had handled our own public relations business in Santa Barbara through City, County and State people and we had chosen not to go the public hearing route. That we had tried to warn L A. Dist. Engr. of Corps of what he faced and *we preferred not to stir the natives up* any more than possible.[20] (Emphasis added.)

When the February 1968 memo was disclosed more than a year later at a United States Senate committee hearing, Santa Barbarans felt they had confirmation of what they already knew; the Interior Department regarded them as "natives" not to be "stirred up," and who could not be trusted with a direct voice in what happens to their community.

With the Interior Department and the Corps of Engineers moving things along effectively for the industry, Union Oil in September requested approval of a third platform near where the first two were to be installed. Why not? responded the Corps.

"With a mighty splash, a 20-story framework of huge steel pipes has plunged into the Pacific off Santa Barbara, signalling the start of a multibillion dollar payoff on a 603 million dollar gamble," the *News-Press* announced in September 1968 as Platform A fell into place.[21]

As the massive structures invaded the channel, county officials wanted to know how far this trend was likely to go. County Supervisor George Clyde asked federal and oil industry executives how many more platforms to expect. At least eight, they replied. "We're sensitive to your concern over the number of platforms, but economics forces this number," said Arthur Joens of the Western Oil and Gas Association. It sounded ominous to county officials, but the forecast was in fact understated.

In November Phillips Petroleum indicated it would erect a second platform in the original federal tract, and the next month Sunray DX disclosed plans for a platform in the same geological line known as the Rincon Trend paralleling the coast less than six miles offshore.

On a slow Saturday night at the *News-Press*, I began writing a series of articles on ten months' observation.

> It will take some doing to pollute the Santa Barbara Channel. It is 1,800 square miles and about 2,000 feet deep at the center. It is flushed by a constant three-to-five knot current running west to east. But man is now capable of despoiling huge masses of land, sea and air simply by industrialization and procreation.

Foremost in my mind was a tanker or freighter slamming into an oil rig or platform, but scores of other potential mishaps were also described in the three-part series. There was no mention of blowouts, only "major spills." I wrote in the concluding article:

> It's late and many concerned citizens have a sinking feeling that oil drilling activity in the Santa Barbara Channel, with all its potential for pollution, is running ahead of the progress made in coping with the possible damage.

The industry was not losing any time. Union's drilling contractor was busy on Platform A in January, having already completed four wells. On the fifth well, Union requested the Geological Survey to waive the rules to permit the driller to put casing down to a far lesser depth than required by the federal regulations. Casing is the outer wall of the well that is intended to keep things like oil and gas from escaping into geological formations and beyond the control of the drillers.

Donald Solanas, head of the Geological Survey in Southern California, was not one to obstruct the oil industry. He approved casing to 239 feet instead of the standard 880 feet. This might have been of no consequence elsewhere. But the bottom of the Santa Barbara Channel is shattered with faults and profuse with shallow oil and gas deposits with only fragile caprock, and in some cases with none at all. That is why the channel has some of the world's most prolific natural oil seeps.

There was nothing new about this condition. The USGS noted later that when Union Oil was drilling in the area in February 1968, oil and gas was seen coming to the surface. The USGS indicated this was not unusual along the Rincon Trend, but also acknowledged that "no documented observation identified the Dos Cuadras sector as a seepage source until work was undertaken ...following the lease sale." Dos Cuadras ("Two Squares") was the name given to the two lease tracts where Union was drilling.

In any case, Union drilled four wells in January 1969 from Platform A, and had asked the Geological Survey to waive casing requirements for the fifth. Drilling began on that well before dawn January 14, and reached beyond three thousand feet when the crew was told on the morning of January 28 to withdraw the drill pipe to run logs to evaluate subsurface geology. Late that morning the well roared back out of control.

Chapter Four

The Awakening: 1969-1970

About 9:30 A.M. Wednesday, January 29, the phone rang at my desk at the *News-Press*. "The ocean is boiling around Platform A," a male voice said. The anonymous caller stated the drill bit struck a high pressure gas pocket which kicked backed out of control and everyone except the drilling crew was evacuated from the platform. That was twenty-four hours ago, and during the night the gas eruption turned into an uncontrolled rush of crude oil. "Thousands of tons of oil are headed for the beach," he said. An exaggeration? Not in the long run. The caller, who declined to identify himself, said nothing more. He probably had been on the platform, or on a work boat that carried men and supplies to the platform six miles off Santa Barbara.

My news story could not be based on assumptions or anonymous phone calls, so after advising the city desk that something big might be brewing, I began phoning around. City and county officials had heard nothing about a spill or blowout, and no one at the local Coast Guard office was available to talk. Minutes before an extended first-edition deadline, I reached the Coast Guard district office in Long Beach. It verified that there had been a "major spill." Now I could use my anonymous caller's information in a quick seven-paragraph story for the first edition.

"A major oil spill has occurred from Union Oil's platform A six miles offshore from Santa Barbara in the Santa Barbara Channel," the lead paragraph began. If, as many would say later, the blowout was the catalyst for what would become known as the environmental movement, these were the first words published about it. Not exactly words to live by. But during the three hours before the home-edition deadline, I collected more information and wrote:

> Union Oil Co. touched off major contamination in the Santa Barbara Channel yesterday from an offshore platform that is so new it has yet to produce its first barrel of commercial oil.

By the 1:00 P.M. home-edition deadline we had reporters at sea and in the air, and comments by the local Coast Guard commander and enraged local officials. The assurances of safe and clean oil production did not hold long enough to produce a barrel of oil. And when disaster struck, the federal government and the industry withheld information as long as one whistle-blower would let them.

Events of the following days, weeks, and months came rapidly and chaotically. It would take a book to relate them. Indeed, several books have been written just about the blowout.[1] This chapter attempts to convey the sense of chaos, outrage, and frustration of the first two years after the event that one writer called "the blowout heard 'round the world."

The morning of the blowout was followed by an afternoon of coverup, a long night of anxiety among those who knew the truth, the next-morning tip by a whistle-blower, an afternoon of panic among local officials, and a late-night meeting called by Lieutenant George Brown, Coast Guard group commander at Santa Barbara.

Under the National Contingency Plan in place at the time, the 27-year-old low-ranking officer became on-scene commander to cope with the spill which soon became a massive environmental disaster. Lieutenant Brown faced irate local officials, perplexed state and federal agency representatives, news reporters and subdued Union Oil executives. In full uniform, including rows of ribbons, Lieutenant Brown announced that Union Oil was responsible for plugging the well and cleaning up the spilled oil. He had delayed reporting the incident, he said, because the seriousness of the situation was not apparent until late on January 28, and it was difficult to reach officials after business hours. Union Oil executives, playing down the whole matter, asserted that the leakage would be controlled soon.

The blowout drew unusual nationwide and international attention not only because of its magnitude and threat to marine life and recreational resources, but because it happened off Santa Barbara, legendary seaside paradise. The community's scenic setting and benign climate were widely celebrated. Only slightly less admired worldwide were its profusion of natural resources—other than oil.

Nevertheless, the national press wearied of the event after a few weeks. On the scene, however, the drama continued with new episodes daily. Even when the well was plugged after eleven days, oil continued to surge to the surface. Once control was lost below the truncated casing, oil escaped upward outside the well through acres of fractured ocean bottom around the platform. Little could be done beyond covering the beaches and nearshore waters with straw which absorbed some oil which could then be scooped up and carted off to waste dumps. The crews ran out of waste dumps before they ran out of straw.

Nor was the January blowout the last. On February 24, another well blew on the same platform.[2] But Union Oil and the Coast Guard announced only that more oil was leaking under the platform. Nobody needed the announcement. Beach dwellers along the coast were afflicted with a new, thicker layer. The *News-Press* published a daily box announcing how many days the leak had continued, and asking, "When Will it End?" The box ran throughout the year, and ten days before Christmas there was assurance that the end, indeed, was not near. A pipeline had ruptured, again on Platform A, spilling tens of thousands of gallons into already contaminated waters. This was followed five days later by a new leak or a surge from the previous one. On December 20, the box score read: "Oil Spill in its 327th Day: When Will It End?"

Washington Wakeup

The stage for the disaster had been set by the Democratic administration of Lyndon Johnson, but the catastrophic events awaited the inauguration of Richard Nixon. He took office barely a week before the initial blowout. Much to the consternation of environmentalists, Nixon appointed Governor Walter Hickel of Alaska, friend of land and oil developers, as his Secretary of the Interior.

Hickel had been characterized variously "as a land exploiter, a champion of business interests over public interests, a foe of the Eskimos, Indians and Aleuts, and a man who worries little about air or water pollution," the *Los Angeles Times* reported before the confirmation hearings.[3]

Hickel's contentious Senate confirmation came only four days before the event that would occupy his attention for much of his abbreviated tenure. On February 2 Hickel flew to Santa Barbara, followed by the president in March. The president appointed two White House committees to study the incident. Senate subcommittees met in Santa Barbara and Washington to do the same. Alan Cranston and John Tunney, representing California in the United States Senate, flew in. Admiral Willard Smith, commandant of the Coast Guard, followed. Much of the Interior Department's bureaucracy came for a first-hand view.

Weary and harried when he arrived on the night of February 2, Hickel was exhausted and shaken when he departed the next afternoon. He was shocked by the scene during a flight over the channel. "The pollution is much more severe than I had anticipated," he said.

On January 31 the Coast Guard estimated the slick to be two hundred square miles, but Dick Smith of the *News-Press*, after a flyover, said it was much greater. Fred Hartley, Union Oil president who also arrived on the scene, said "the flow at the present time is predominantly gas." It was typical of the contradictory assessments to come, and led to decisions that were changed with each passing day.

Hickel initially requested the oil companies to stop drilling voluntarily, then soon ordered them to stop. "I did not have the authority to close down the drilling rigs because they were polluting the water," Hickel wrote later, "but I did have the authority to close them down if they were wasting the oil—which they were. What a sad commentary on the attitudes toward the environment I inherited as Secretary!"[4]

Amid the turmoil, Hickel's requests and orders on drilling changed frequently. The secretary had to decide whether the leak would be better controlled by stopping drilling from Platform A or attempting to relieve downhole pressure by keeping the wells open and operating.

In March he used his authority to add a 34,000-acre buffer zone to the 21,000-acre federal ecological preserve that the Johnson administration had established in 1968. These zones are seaward of the state sanctuary that runs along the coast thirteen miles from Summerland to Coal Oil Point. He stated that no oil leasing should have been permitted near the ecological preserve fronting on Santa Barbara. There were, he said, more suitable places to produce oil.[5]

During his stay in Santa Barbara, Hickel did not blame Union Oil for the blowout. Federal rules had not been violated, he said; they were simply ineffective. Nor did he chastize Santa Barbarans—

many of whom waved clenched fists—for protesting. "As early as 1955 the people of Santa Barbara had fought against the establishment of offshore rigs on their resort doorstep," he wrote later. "I believed they had the right to yell as loudly as they wanted, because they had yelled before the fact."[6] (As we have seen, they yelled long before 1955, but Hickel had the right idea.) Stewart Udall, Hickel's predecessor, also repeatedly stated after the fact that leasing the channel was a mistake. He called it his "Environmental Bay of Pigs."

Hickel promptly rewrote some regulations to enforce the Outer Continental Shelf Lands Act. He tightened up requirements for casing and blowout preventors, and—over the protests of the industry—imposed "absolute liability without cause" on the industry for spill damages[7] As Robert Easton phrased it, "A new Hickel was emerging."

Behind Closed Doors

The White House, too, was looking for answers. Santa Barbarans had thousands of answers, in addition to piercing questions and caustic comments. They had written thousands of letters to the editor, and on February 11 the *News-Press* mailed them to the president.

> Because the *News-Press* believes these letters are an accurate reflection of local public opinion regarding offshore oil exploration in the Santa Barbara Channel we are sending tearsheets of all of them, as well as the daily *News-Press* editorials on the subject, to President Nixon and all elected officials representing this area at the federal and state levels.

The president replied on March 14 to say he had shared the letters and editorials with Hickel, and that they were both very concerned. "You may be sure," Nixon said in his letter to executive editor Paul Veblen, "that this Administration will do everything possible to protect America's great natural resources."

Maybe the president *was* concerned; one week later he descended via helicopter on a pristine beach just west of the harbor to observe the community's plight. Workers had toiled many hours to sanitize the scene. But a couple hundred yards away, behind restraining cordons, a crowd chanted, "Get Oil Out! Get Oil Out!" The beach was the first in Santa Barbara to be smothered a month and a half earlier by the capricious slick. On the afternoon of

February 4, the sound of breakers suddenly ceased, and a silent swell came ashore under an undulating layer of oil. As the swell receded, the oil settled as a carpet on the sand. Each succeeding swell brought another layer of crude.

Here, then, stood President Nixon six weeks later to tell the press and local VIPs that "the Santa Barbara incident has frankly touched the conscience of the American people." He said he would consider the local demand for a total and permanent ban on drilling and production in the channel. In any case he felt that new regulations drafted by Hickel would assure safe and clean operations in the future.

The president had already appointed a committee to investigate the problem. It was headed by John Calhoun, vice president of Texas A&M University and former science advisor to the Interior Department. The panel met February 19 at the University of California's Santa Barbara campus—behind closed doors. According to Dr. Calhoun:

> We are not here as an investigating team and we are not here to find out who did what and why, except we would like to know what did the Santa Barbara people experience—what did they learn that they think is significant for later situations.

Locking the public out of the discussion was a strange way to discover "what did the Santa Barbara people experience." But Dr. Lee A. DuBridge, Nixon's science advisor, had instructed the panel to do more than that. He told the committee to look for ways to produce the oil without degrading the coastal communities too badly, and for ways to restore them if the worst happens.

News-Press editors decided to thrust me into a confrontation by protesting being barred from the meeting. As it turned out, the panel fell behind schedule and even the post-meeting briefing was cut short by Dr. John Steinhart, an assistant to Dr. DuBridge.With my tape-recorder running, I objected:

> We (the press) are already barred from sitting in on the sessions. Now [you] are cutting these press sessions, which are spoon-fed conferences, in addition to our being barred from the deliberations. The people of this community are already incensed at the federal government for permitting this to happen in their channel. Now the federal government comes here to investigate this thing—a panel representing the White House—and again the people are told they have no business sitting in on these deliberations. How do we explain this to our readers?

Dr. Calhoun replied:

> Gentlemen, I think the first thing you should explain and make clear is that we are not here to investigate the Santa Barbara incident. We are here as a group of scientists and engineers to give our best overall scientific thinking to advising the President's science adviser on the positions he would take or should recommend to the Congress and to the President.

I continued. "This does not respond to my question. Why is the public not permitted to sit in on its own business?" Dr. Steinhart responded:

> I can answer that one, I think. White House advisory panels have never to my knowledge at any time—the press does not sit in on cabinet meetings nor on deliberations in the State Department among experts trying to get an idea of what to do. I think you can appreciate why.

"No," I said, "that is the question I have just asked." Dr. Steinhart tried again:

> There are state officials who have been active in this—have been participating all morning, in fact, with us, and I have no doubt that they may participate further. This is not a public hearing....But there is a full scale investigation going on [by another agency]. If we really wanted to investigate what happened here, we would have to, for one thing, have a different kind of panel, and for another, we'd have to go at it with some regard for due process and all that sort of thing....We are specifically directed not to consider that.

Then Dr. Calhoun added:

> To get value out of discussion among a group of scientists and engineers, they have to interact and they have to bat their ideas back and forth. This is a sifting process of scientific and technical ideas. It's just like a conversation among experts in any field—what are the angles? This is what we're engaged in, really.

(In the book that Dr. Steinhart later wrote with his wife, Carol, he essentially repeated the rationale for closed meetings.[8]) In its report eight months later, the committee recommended public hearings and greater public participation in offshore oil leasing decisions. Although it had insisted on meeting without public attendance, the committee apparently felt that open meetings were a good practice for other groups. "There is no satisfactory substitute for hearing a variety of views in a public forum," the committee said. It also suggested closer federal consultation with local government and greater consideration of local interests. This conversion to openness in government was made in secret. That was the only way, we were told earlier, that it could be done. Allowing the people to participate would have spoiled everything.

Beyond that the committee suggested greater efforts to reduce the likelihood of spills, to advance cleanup technology, and to declare some sensitive but potential oil-producing areas off limits, at least for the time being.

In May, a *second* White House panel of scientific and technical experts met—secluded, of course—in Inglewood, to ponder "the future operation of the Union lease. . . . [T]he publicity and information releases from the panel must be strictly controlled," the panel was instructed by William Ruckelshaus, then assistant attorney general. "An adverse report from a blue-ribbon panel such as this could be very detrimental in the pending and anticipated litigations."[9] The public again wasn't being let in on it's own business.

This time the resulting recommendations included continuing production of the oil-bearing formations under Platform A until they were depleted. *That* should stop the leak. But in how many years? In the meantime, Union Oil and its three partners would be punished with even greater riches. Public response was immediate, overwhelming, and hostile, as we shall see.

Congress Stirs, Sort of

Congress, too, was stirring with indignation over what the industry had done to Santa Barbara. Or so it seemed. Eight days after the blowout, the Senate Public Works Subcommittee on Air and Water Pollution took testimony in Washington, and before the end of the month it heard two more days of comment in Santa Barbara. In May the Senate Interior Subcommittee on Minerals, Materials and Fuels sat for two more days of testimony on a bill by Senator Alan Cranston to ban further drilling in the channel. He was not the only one drafting legislation. By March 15, there were forty-four pollution bills in Congress, mostly dealing with offshore oil activity. They included bills to buy back the channel leases, to make the entire channel an oil-free sanctuary, and to ban or delay further drilling.

Just before the Christmas recess, Congress passed the National Environmental Policy Act (NEPA), the nation's basic resource protection law, but nothing was enacted specifically to protect the channel. Congress was not taken with Santa Barbara's oil problems. Returning from the Senate subcommittee hearing in May, I wrote:

> The impression left by two long days of testimony on the Senate bill [to ban drilling] is that there is little understanding of or sympathy with the point Santa Barbara is trying to make.
>
> Several senators appeared to feel that it would be discriminatory to single out Santa Barbara for protection from platforms and pollution, while permitting oil exploration elsewhere off the U. S. coast.
>
> "Why, they wanted to know, is Santa Barbara any different?

The state legislature too refused to support any bill that would seriously curtail oil activities in the channel. After the Assembly Finance Committee soundly defeated a bill to ban further drilling in state waters, Jesse Unruh, the bill's author, concluded that the committee was "in the pocket of the oil companies." There was little hope for congressional action if the California legislature was not interested.

Administrative agencies seemed to be less influenced than legislators by oil interests that help finance election campaigns. The State Lands Commission suspended offshore drilling in California waters a month after the well blew out in the federal lease. Production could continue, but new drilling was suspended. In April the commission deferred a request by Standard Oil of California to drill new wells from Platform Hope. On July 31, the day after industry lobbyists succeeded in killing the Unruh bill, the issue was again before the State Lands Commission for administration action. Its executive officer, Frank Hortig, wanted the Lands Commission's ban lifted because the state was losing production royalties.

Caspar Weinberger, who later as Secretary of Defense was prepared to launch the doomsday bomb, appeared terrified at permitting more drilling when he saw what could happen without appropriate safeguards. As Governor Ronald Reagan's finance director, Weinberger was an ex-officio member of the Lands Commission, together with the lieutenant governor and the state controller who shared Weinberger's views. It would be some months before the commission was ready to lift the ban, and then only on a case-by-case basis.

Appeal to the Courts

In seeking relief, Santa Barbara county, the state and some cities and residents looked also to the courts. Plaintiffs soon discovered how far the oil industry's influence extended.

Ten days after the spill, $1.3 billion was demanded of Union and its partners in a class action suit on behalf of anybody who suffered damages. On February 21, the state of California filed a $560 million damage suit on behalf of the state, the city and county of Santa Barbara, and the city of Carpinteria. To press his case alleging negligent oil drilling, Charles A. O'Brien, chief deputy attorney general, cast about for expert witnesses. They all ran for cover. Most of the prospects were working for oil companies or were on university faculties that enjoyed oil industry grants. Many oil industry executives sat on university governing boards.[10] Charles O'Brien noted:

> The university experts all seem to be working on grants from the oil industry....There is an atmosphere of fear. The experts are afraid that if they assist us in our case on behalf of the people of California, they will lose their oil industry grants....We recognize that these petroleum engineering experts have only one market for their expertise: The petroleum industry. We would hope that the petroleum industry would recognize the shadow it casts over university research when it creates a climate in which university experts are afraid to use their talents on behalf of the people this state.[11]

O'Brien made a public issue of the matter through press conferences in which he also accused most regulatory agencies of being in bed with the companies they were supposed to regulate. Eventually he assembled a team of expert witnesses that included Dr. Paul Witherspoon, professor of petroleum engineering at the University of California at Berkeley; Dr. Harry Brandt, professor of mechanical engineering at the University of California at Davis; and Robert Sharp of Santa Barbara, a retired oil drilling engineer of extensive experience and unchallenged credentials. Marvin Levine, deputy county counsel, worked closely with O'Brien in preparing and presenting the case.

In another suit, Levine asked the federal court to enjoin further leasing and drilling in the channel without public hearings and without the right to inspect information on which oil activity permits were based. In an unusual twist, the American Civil Liberties Union of Southern California filed a similar suit on behalf of seventeen Santa Barbara residents. A. L. Wirin and Fred Oakrand of the ACLU argued that a clean environment is a basic right. These approaches were new to the courts, and the judges did not know what to do with them except throw them out.

"[T]his matter of protecting the environment from pollution had a strange and foreign ring to the ears of the court," Levine said.[12] "The courts didn't know what we were talking about. The Outer Continental Shelf is government land, the judges said, and the federal government had a right to make decisions about what to do with its own land."

During this frustrating process, Levine wrote a memo—to himself—in which he questioned the right of one generation to take from generations yet unborn, who cannot be heard now, what should be theirs. They are being deprived of due process, Levine wrote. "Should not such persons be given adequate representation *now* and be heard from *now* since it is *their* life, liberty and property which we seek to deprive them of forever?" Levine asked.

He was elated to find that the National Environmental Policy Act, enacted in December, required the federal government to "fulfill the responsibilities of each generation as trustee of the environment for succeeding generations."

> We lost in the district court, and while we were on appeal, the National Environmental Policy Act became law. . .NEPA changed it all. It gave us what we wanted—notices, public hearings and environmental considerations. What we couldn't do judicially was accomplished legislatively.[13]

NEPA became effective January 1, 1970, a few weeks too late to impede Sun Oil Company from installing Platform Hillhouse near where the well blew out nearly a year earlier. The platform, already approved by the Interior Department, received a Corps of Engineers permit November 7, and Sun Oil was ready to barge it from a shipyard in Oakland. Opposition mobilized on two fronts. Get Oil Out organized a "fishing" fleet that would be strategically anchored to prevent the barge from delivering the platform to its intended location. Meanwhile, Levine and Wirin, having lost everywhere else in the judicial system, went to Washington to seek an injunction from the Supreme Court.

On November 20, the twenty-story high steel platform framework approached the "fishing" fleet which, at anchor, had the right-of-way over a vessel under way. Donald Solanas, the ranking Interior Department man on the scene, could do no more than threaten to report the "fishermen" to the Department of Justice.

Across the continent, Levine said later, he and Wirin were supplying bourbon and hand-written pages of an evolving brief to a court reporter who required constant stimulant to continue typing. On the "western front," the "fishing" fleet, now down to one vessel, held its position but it was just far enough off the mark to permit the tug and barge to dump the structure on target. As it tipped off the barge on its side, it was supposed to right itself by tilting bottom-first to the seafloor. It did not work that way. It settled in upside down, with its huge leg bottoms pointing skyward several feet above the water. A derrick barge was called in to tow the structure to deeper water where, with cranes and cables, it was righted and towed back to its intended location. This took several days which Levine and Wirin used to press for a Supreme Court injunction against installing the unwieldy framework. The court ruled against the county on November 25, and the structure settled into place the next day. It ended a dramatic battle whose outcome might easily have been predicted, but no one involved regretted having made the effort.

Injunctive relief efforts, then, were largely dead by the end of 1969, but the damage suits carried over into 1970 and beyond. In addition to suits filed by the state, county and two cities, and the class-action suits, several private damage suits were filed, and the district attorney had initiated several unsuccessful anti-pollution actions against the oil industry.

Protests, Official and Grass Roots

Seldom have local governments and community protesters been in such agreement. The blowout immediately spawned Get Oil Out. Within twenty-four hours the organization was on its way under the leadership of Alvin Weingand, a former state senator; James (Bud) Bottoms, an artist and sculptor; Marvin Stuart, a public relations specialist; Lois Sidenberg, longtime active conservationist, and others. Before mid-year it had 70,000 signatures on a petition demanding an end to oil operations in the channel, and had 150,000 signatures by the end of the year. (It grew to 200,000 by April 1970.) GOO organized public rallies, testified at hearings, petitioned President Nixon to curb the oil industry in the channel, and applied its influence on the legislature and Congress.

GOO is not the only resource protection group that owes its existence to the blowout. A legal organization immediately emerged that evolved into the Santa Barbara Environmental Defense Center, a highly professional and effective public interest law firm. The January 28 Committee became the Community Environmental Council which today is an ecological think (and action) tank of international repute.

Public protest got strong support and direction from the Sierra Club, largely under the leadership of Fred Eissler, a Santa Barbaran who held a seat on the national organization's board of directors. The League of Women Voters of Santa Barbara eagerly plunged into the sticky issues of offshore oil development. Its position was never in doubt.

Among county officials, Supervisor George Clyde led the attack, starting in 1968 when he headed the Santa Barbara Channel Oil Advisory Committee. The Santa Barbara City Council appointed an Environmental Quality Advisory Board (EQAB) which was active during periods of good leadership, and waned at times under lesser direction.

With these and other forces at work, protest was manifested daily throughout the year: petitions, rallies, public hearings, news conferences, picketing, thousands of letters to the editor, visiting celebrities adding their (sometimes uninformed) voices; marches with strident placards; teach-ins and numerous other activities.

In addition to the fish-in, one community protest must be noted: the Easter Sunday demonstration on Stearns Wharf.

The April 6 event started as a waterfront park rally, organized by a group without benefit of an organization or name. It brought out four hundred or five hundred, including Episcopal Bishop C. Edward Crowther who said, "We are in fact protesting man's right to a decent sort of life. This I believe is a fundamental religious preoccupation." The last speaker suggested a march on Stearns

Wharf, the downtown pier that was owned by the city and leased to a group that enjoyed revenue from wharf restaurants as well as fees from oil companies who used the pier to store and transfer equipment and supplies for their offshore operations. The lease was to end in 1973, but the crowd wanted the city to take back the wharf immediately and evict the oil industry.

Almost everyone at the rally walked the short distance onto the wharf where they signed petitions protesting industry's use of the pier, and agreed to ask the city council the next day to terminate the lease. As the protestors began leaving the wharf, they were confronted by two large trucks carrying heavy oilfield equipment onto the pier. The crowd refused to let them through, and about a dozen sat down in front of the first truck. Three police officers moved in, and more protestors sat down until about forty people were settled on the wharf planks. Here was middle-age, middle-class Santa Barbara sitting in front of the gigantic trucks to defy the multi-national oil industry. It was not planned or organized; it was a spontaneous radical response by Santa Barbara's "respectables," including university professors.

The police officers stepped aside to confer, and the crowd began singing "We Shall Get Oil Out," to the tune, of course, of "We Shall Overcome." The officers talked to the driver, then to the crowd which responded by chanting, "More pay for cops! More pay for cops!" The officers stepped back, the driver turned off his engine, and the crowd cheered.

One motorcycle officer was applauded as he moved a row of protesters back to open a lane for passenger cars. He then turned around and moved the crowd back on the other side. "I came back for an encore," he said.

Police Lieutenant Harry Hobson conferred again with the truck drivers. They started their engines, and began backing off. The crowd roared. Later, of course, the trucks returned. But for an hour and a half the anti-oil movement prevailed in a dramatic demonstration.

The Press Responds

As long as the press remained interested, it played an important role. But editors have short attention spans. *News-Press* editors, of course, were interested as long as oil was leaking, and they were distressed over news outlets that dropped the story so quickly.

By spring "the public by and large had been given the impression that the leak at Platform A had been repaired, and that except for some cleanup operations everything was under control," wrote David Snell in a memo to his managing editor at *Life* magazine. He produced clips from the *News-Press* which indicated otherwise, and got permission to take a photographer with him to Santa Barbara to do a story.

The resulting June 13 article, "Iridescent Gift of Death," lit a short fuse in the oil industry and the Interior Department. They were particularly irate over a photo-story of conditions at the famous pinniped rookery on San Miguel Island, forty-five miles southwest of Santa Barbara.

> Until we became weary and sick of the tally, we counted over a hundred dead sea lions and elephant seals in the immediate area....most were new-born pups....Here and there we came upon oil-drenched pups and that cried weakly and thrashed about like scalded rats, their eyelids gummed shut, umbilicals stained and caked.

Oil industry and Interior spokespersons protested that the number of dead was not unusual during the pupping season, that whatever oil was present came from natural seeps, and that neither Snell nor the people with him were qualified to assess the biological damage. Snell wrote an in-house memo:

> In all my years as a professional journalist at home and overseas (and never, I might add, in behalf of any wildlife or conservationist group), I have yet to see anything that would remotely compare with the efforts by Union Oil and its allies to bottle up the truth about Santa Barbara's spill and its appalling effects.

Snell did not visit the island alone. With him were Paul Veblen, *News-Press* executive editor; Dick Smith, amateur naturalist and author of three books on wildlife; Smith's daughter Judy; Zoologist Jodi Bennett; and a retired Navy captain, Eric W. Pollard. Two weeks earlier Ian McMillan had visited the island for an article for *Defenders of Wildlife News*. McMillan was a rancher-naturalist, lacking advanced academic degrees but noted as an author on wildlife conservation.

The opposition recruited people of impressive credentials to refute Snell's observations. But Veblen pointed out that they were selected by Union Oil for a brief Union-sponsored trip to the island. Veblen wrote to Snell: "As for lay observers, including you and me, we know what we saw."

The *News-Press* printed what it saw, and was honored for it. The California Press Association named Stuart S. Taylor "publisher of the year" for the *News-Press* oil coverage;[14] the Associated Press Managing Editors Association cited the *News-Press* for its cooperation in AP coverage of the oil story; and I was awarded second place in the nationwide Scripps-Howard Foundation's Edward J. Meeman Awards competition for conservation writing on the basis of my November 1968 articles anticipating by two months a catastrophic oil spill.

The Impact

Measuring the impact of the spill in 1969 was—and is—a futile pursuit. How much oil was spilled, how much did it damage marine wildlife, what was the impact on local economy? The industry and federal government minimized the consequences. Oil companies maintained that the damage was temporary, and anyway, natural seeps pollute the offshore waters daily. Damage *was* temporary. Breaking a leg, arm or rib is temporary, and the fracture eventually heals. But it is also traumatic, painful, costly, grossly inconvenient, a serious setback in one's intended progress, and an experience one diligently strives to avoid repeating.

How much oil spilled from Platform A? Because it came up through oceanbottom fissures over a vast area and at varying flow rates, there was no way to measure it. You can take your choice among the many estimates. On the low side, the Geological Survey set the figure at 18,500 barrels. A team headed by Dr. Michael Neushul, a biological scientist on the faculty of the University of California at Santa Barbara, set the high mark of 780,000 barrels. Alan A. Allen, a scientist at General Research Corporation in Santa Barbara, put the number at 80,000 barrels. He estimated the thickness of the oil on the water by its various colors as determined from earlier spills, and by the area of the slick from time to time. The Coast Guard selected the round number of 100,000 barrels. (A barrel of oil is forty-two gallons.)

How much it damaged resources is also in dispute. As far as anyone knows, all species recovered, although some were severely depleted for a time. With too little information on what plants and animals existed before the spill, there was no way to accurately estimate how the spill had changed things. And some attempts were criticized because they were made by researchers selected and paid for by the oil industry. Dale Straughan, a marine biologist sponsored by the Western Oil and Gas Association, reported minimum damage to coastal creatures. Her report was attacked by Dr. Max Blumer, a senior scientist at the Woods Hole Oceanographic Institute, who called her techniques sloppy and her conclusions not supported by the evidence. (More on this later.) Commercial fishers found fishing grounds seriously disrupted once the oil cleared enough to permit the use of their boats. Eventually the fisheries returned to normal.

How many birds and fish perished? You could count the ones that washed ashore, but not those that sank, were buried, washed out to sea, rotted unnoticed, or disappeared in other ways. The California Department of Fish and Game reported it knew of 3,600 birds that died. More than 1,500 oiled birds were treated at a waterfront station, but only 168 survived. The best that can be said for the effort is that much was learned for treating birds in subsequent spills.

The economic loss was no easier to calculate. Tourism revenue fell in 1969, of course, and recovered in succeeding years. Waterfront property damage was tabulated in a class action lawsuit that later awarded nearly $6.5 million (in 1969 dollars) to owners of beachfront homes, apartments, hotels, and motels. Commercial and recreational boat-owners and nautical suppliers collected $1.3 million for damage to property and loss of revenue, but the court would not award damages for longterm harm to the fisheries.

The suit by the state, the city and county of Santa Barbara, and the city of Carpinteria against Union Oil and its partners was not concluded until 1976. Just before it was scheduled for trial, the oil companies settled for just under $9.5 million, again in 1969 dollars. The court approved damages for loss of property, but not for loss of public revenue, which was substantial. No plaintiffs felt the awards were adequate, but they got what they could. This in no way was a measure of the real damages.

The hour-by-hour events made 1969 a tumultuous year in Santa Barbara. And yet, it was only the beginning. To fill out the year without a day's respite, Union Oil announced on December 31 that it planned to install a third platform off Santa Barbara "as soon as possible." But the Interior Department indicated in mid-January that Santa Barbarans probably would not see the platform for a long time.

Platforms A and B were in place, and Platform C was built and approved for installation before the blowout. But Hickel suspended the permit after the eruption. As it turned out, the twenty-story tower languished on its side in a Vancouver, British Columbia shipyard for seven years while it rusted, and weeds grew tall among the steel beams.

1970 — A Year of Protest

Although it put Platform C on hold, the Interior Department encouraged Union Oil to increase production from Platforms A and B. The president's scientific panel said this would reduce pressure in the subsea oil zones and slow the flow of oil through the fractured ocean bottom. Whether this was well-founded science or a device to enhance private oil profits and public royalties was widely speculated. William Pecora, Director of the Geological Survey, said his office would produce a report that would convince skeptics that this was science, not corporate interest.

When the 77-page document was released January 8, 1970, critics said it was a good report, but not what was promised. "The report. . .draws no firm conclusions as to what caused the leakage," wrote the *Los Angeles Times*. "Nor does it speculate on how the blowout could have been prevented or what the chances are that it will happen again."[15] Most members of the Santa Barbara Environmental Quality Advisory Board were impressed by the report, but noted it did not explain how continued oil production could slow the leakage. The USGS *Professional Paper* 679 described the geology of the channel, included a chronology of the blowout events, the history of oil activity in the channel, and geological and oil activity maps.

At the January 26 EQAB hearing, the Geological Survey submitted the "raw data" on which the oil pumping decision was based, and Union Oil turned over data which had influenced that decision.

EQAB, an advisory board of scientists, engineers and economists, reviewed the data, and eventually not only supported continued production, but suggested that another platform might be installed to help reduce subsea pressure and oil leakage into the channel waters. This incensed Get Oil Out and Marvin Levine of the county counsel's office. They charged that EQAB's information came almost entirely from the Geological Survey and Union Oil who were more interested in oil production than in resource protection. They felt EQAB was undermining a community effort to curb oil activity and clean up the coastal environment.

To help convey this community vision to Washington, a delegation headed by Phil Berry, president of the Sierra Club, crossed the continent January 20 to see Secretary Hickel. I accepted the group's invitation to come along. As we entered the secretary's office, an aide held out his arm to block my way. The secretary, he said, did not want any news reporters at the meeting. I said this was not the secretary's meeting; it was arranged by the Santa Barbara delegation which had invited me to attend. I was pushed back to the outer office where I created a ruckus with Alex Troffey, Hickel's press officer. "You must really hate yourselves to keep me out!" I shouted. "You'll get a lot worse press than if you had let me in." I made myself heard through the double doors between the outer office and Hickel's inner sanctum. Before the afternoon was over I was scheduled for an exclusive interview with the secretary the next day.

During the interview, Hickel suggested an oil-free sanctuary in federal waters extending south about twenty miles from the state sanctuary across the channel to Santa Cruz Island. Thirteen miles wide, it would coincide with the east-west boundaries of the state sanctuary, and would be off limits to future leasing. (Pecora, who sat in on the interview, said later that the sanctuary idea had originated with John Crowell, professor of geological sciences at the University of California at Santa Barbara.)

The plan, disclosed publicly for the first time in the interview, became the crux of the administration's policy through a series of congressional hearings in 1970. But it enjoyed the support of almost nobody except the White House and Interior Department. The oil industry indicated it was too restrictive. Santa Barbarans said the proposed sanctuary was too small to do much good because the oil industry was not much interested in the sanctuary area anyway, and the rest of the channel would remain open for leasing. Twenty leases within the proposed sanctuary would be taken back. The fifty-two lease tracts outside the sanctuary could be exploited by the oil companies. It was not what Santa Barbarans had asked for, but, some agreed, it was a start. Even Congressman Charles Teague said he hoped for amendments that would place more limits on the oil companies.

First Anniversary

Hickel's proposal was not the first indication in early 1970 that Santa Barbara's protests were being heard. President Nixon and Governor Reagan announced great plans for dealing with pollution. Things were looking up, then, as seventeen organizations joined with the January 28 Committee to plan for the first anniversary of the blowout. Under the leadership of Marc McGinnes, a 28-year-old lawyer, the event at Santa Barbara City College would be on a bluff overlooking the channel with the oil platforms clearly in view.

The group booked twenty-seven speakers from Congress, the state legislature, university faculties, labor unions, student organizations, the state administration, and from leaders of national environmental, political and health organizations. More than 1,000 turned out, overflowing the 800-seat auditorium. Loudspeakers accommodated the overflow outside.

Despite promises of new approaches by Washington and Sacramento, speakers at the conference were wary, calling for action at the bottom to induce serious movement at the top. The keynoter was Dr. Paul Ehrlich who had just become famous for his book, *The Population Bomb*.

"The rising talk of environmental concern is designed to convince concerned citizens that something is being done, while in fact we are going down the same old path," Ehrlich said. Population control is required to solve other environmental problems, he said, "but no government in the world is doing a thing about it."

Stewart Udall agreed. He said the usual proposals to end pollution never seem to acknowledge the overriding problem of rapid population growth. There is more to saving the planet than stopping oil pollution, he said. Although he was the Interior secretary who had approved the 1968 oil lease sale in the channel, he was popular in Santa Barbara because he acknowledged that he had made a mistake. Now he asked his large and sympathetic Santa Barbara audience: "Are you going to save the orchards, and a little open space? Are you willing to ride a bike or walk to work?"

Another hero of the conference was John Burton, a member of the California legislature who was later elected to Congress. Thunderous approval greeted his proposal to nationalize the oil industry. "We have to eliminate the profit motive that has induced industry to make the most money as fast as it can and the public be damned," he said as applause nearly drowned out the end of his sentence. "We must use their [oil industry's] profit and intelligence to do something about the problem that they have caused through the years of exploitation of our natural resources. . .The first thing that must be done is to nationalize the oil industry." Heady stuff, and it continued.

The conference adopted "The Santa Barbara Declaration of Environmental Rights," drafted and read by Dr. Roderick Nash, professor of history at UC Santa Barbara, and author of the 1967 classic, *Wilderness and the American Mind*.

"All men have the right to an environment capable of sustaining life and promoting happiness," Nash intoned over national television. "If the accumulated actions of the past become destructive of this right," he continued in when-in-the-course-of human-events cadence, "men now living have the further right to repudiate the past for the benefit of the future. And it is manifest that centuries of careless neglect of the environment have brought mankind to a final crossroads. . .We need an ecological consciousness that recognizes man as member, not master, of the community of living things sharing his environment. . .We propose a revolution in conduct toward an environment which is rising in revolt against us."

Santa Barbara was radicalized, and the ferver was spreading. It was an election year, and from Washington and Sacramento to the local level, politicians launched their campaigns on the conservation issue generally, and many on the Santa Barbara problem specifically. Nixon's first term was not up for two more years, but every congressional seat would be contested in 1970, as would one-third of the Senate seats. Nixon, the staunch conservative who once boasted that Americans consumed vastly more resources per capita than people anywhere else in the world, now quickly crafted a program of natural resource protection. Governor Reagan, up for re-election in 1970, did the same.

Santa Barbara's legislators in Washington and Sacramento were also steadfast conservatives: Congressman Charles Teague, State Senator Robert Lagomarsino, and State Assemblyman Donald MacGillivray. But they too sensed the public mood. They protested that they were as thoroughly "anti-oil" as anyone. They introduced anti-oil bills to prove it. Campaigns were debates over who was the greatest bulwark against offshore oil. Activists, of course, viewed the recently-converted with suspicion or even contempt. But the incumbents' constituencies held. Successfully pre-empting the anti-oil issue from the Democrats, Teague, Lagomarsino, and MacGillivray were re-elected. So, too, was the governor.

The State Lands Commission, meanwhile, continued to resist industry requests to lift the ban on new drilling in the tidelands. For several months the commission even denied permission for a Sun Oil pipeline which would run through state waters from Platform Hillhouse to onshore storage tanks in Ventura County.

Public Protests

A year of public protest started in January when a huge banner was hoisted from a sixteen-foot Boston whaler off San Clemente where President Nixon was relaxing at the seaside "winter White House." The banner, unfurled by James (Bud) Bottoms and Paul Molitor, displayed three letters: GOO. It was part of Get Oil Out's continuing attempt to get the president's attention. If the president did not respond, many others did. By March, GOO's membership had shot up to 3,000, and its petition to phase oil out of the channel now had 200,000 signatures.

Protests became less flamboyant, but no less emphatic. They were now in the courts, at legislative hearings, and in the press. The movement had become more sophisticated, and could argue economics, engineering, science, law, politics, and public procedure as skillfully as the oil industry. All the movement lacked was money and more people in high places with courage and a feeling for what was happening to the planet.

Legislation, But No Laws

Some legislators had a feel for it, or were at least impressed by the number of voters who had that perception. The California legislature was inundated with three hundred environmental bills, many dealing with offshore oil. Virtually none was passed.

The big achievement of the legislature in 1970 was the enactment of the California Environmental Quality Act (CEQA) nearly a year after Nixon had signed the federal National Environmental Policy Act. The two laws were cousins, not twins. The California act was much more demanding in an attempt to protect natural resources, but nobody seemed to know it at the time. Until the California Supreme Court ruled in the *Friends of Mammoth* case in 1972, the state applied the act only to public works projects. The court, however, said that it applied to any private project requiring a public agency permit. That changed everything on the state level, but it did not reach into the Outer Continental Shelf.

In Congress, meanwhile, Senator Frank Moss of Utah was showing more interest—if not sympathy—than anyone outside of California. As chair of the Senate Interior Subcommittee on Minerals, Materials, and Fuels, he scheduled hearings on channel oil legislation in March in Santa Barbara and in July in Washington. The March 13-14 hearing centered on a bill drafted by Senator George Murphy of California and Congressman Charles Teague whose district faced on the channel. Both were conservative Republicans, but they wanted to end all oil activity in the channel except the pumping that was intended to reduce subsea pressure. That would mean buying back nearly all the seventy-one federal leases that were sold in 1968. The money would be raised by permitting commercial production from the Elk Hills Naval Petroleum Reserve. Most of the testimony came from Santa Barbara area residents who wanted the offshore oil industry "to clean up, pack up, and get out," as expressed by Alan Eschenroeder of the Santa Barbara City Council.

When the subcommittee met in July in Washington, the administration was ready with its sanctuary bill. Introduced in June, the bill was already thoroughly denounced by Santa Barbarans associated with GOO. Nixon's plan was essentially what Hickel had suggested in his January interview: A thirteen-mile-wide sanctuary seaward of the state sanctuary and extending across the channel to Santa Cruz Island. Not including, of course, the "pressure-reducing" platforms. The federal government would buy back twenty leases in which the industry had shown slight interest, and leave fifty-two others for exploitation. The large Santa Barbara delegation repeated its "pack up and get out" theme. The industry, on the other hand, felt the Nixon-Hickel plan was unduly restrictive, if not downright illegal. Among the witnesses, only Hickel thought it was a good idea.

The administration's bill got another airing in the House in September before the Interior Subcommittee on Mines and Mining, under Congressman Ed Edmondson of Oklahoma. Like Senator Moss, Edmondson was a mid-continent resident who had to be convinced of the concerns of coastal dwellers. This time Californians got help from a state assemblyman from another coastal city. Pete Wilson, former mayor of San Diego and destined to become governor of California, led a delegation that asked for federal sanctuaries seaward of *all seven* state sanctuaries.

This time Eschenroeder set aside his "pack and get out" proposal and tried a new approach. He and County Supervisor George Clyde urged a moratorium on further oil activity until technology for ocean-bottom production wells could be considered safe. (Later that year when Humble Oil disclosed a plan for oceanbottom production wells at the unprecedented depth of 2,000 feet, the community said go test them somewhere else.)

In the end, neither the legislature nor Congress passed any laws affecting oil in the channel. Congressman Teague dutifully carried the administration's sanctuary bill, but he must have been uncomfortable with the task since he had legislation of his own that went much farther than the Nixon-Hickel proposal in curbing the industry. He said he would reintroduce his bill as soon as the new Congress convened in January 1971.

Another Year in Court

There was no more progress in the courts than in Congress and the legislature. Among the frustrated lawyers was David Minier, Santa Barbara County's eccentric district attorney. He sued the four owners of Platform A for polluting the county's coastal waters and beaches. The oil companies won a federal injunction against Minier's action, alleging that the blowout was in federal waters beyond his legal reach. Minier said the spill contaminated state waters and local beaches, and that gave him all the authority he needed. That dispute was on appeal at year's end.

The only legal contest resolved that year came in June when 106 recreational boat owners settled for $150,325 to compensate for physical damage to the vessels that were trapped when oil oozed into the Santa Barbara marina. But James Oppen, an attorney representing other boat owners, called the settlement "peanuts" and a "sellout....This settlement would cheapen my cases," Oppen told the court.

At this comment, Wallace Odemar, a member of the three-judge special panel hearing the class action suits, perked up. He said he resented the implication that the court would be unable to weigh his cases objectively if the settlement were allowed. "Isn't that a slam at the court?" he asked. "Not yet," Oppen replied.

The next month as Oppen resumed his attacks on other attorneys and hinted that the court was not up to its task, the presiding judge advised him that the court was not interested and struck the comments. In April 1971, the judges ruled that Oppen's clients could not be compensated for the loss of use of their boats—only for physical damage. Six of his clients were awarded $54,515, nine others settled before a trial for $65,000. The award to the six, he said, was a compromise between what he asked and what the oil companies offered.

(One judge referred to Oppen as a "chronic bitcher." Oppen, who owned a cabin cruiser, replied that he would name his next boat *The Chronic Bitcher*, which he did. One day, alone aboard the vessel in the channel, he sent out distress signals. He was shortly found dead in *The Chronic Bitcher*. How he died remains a mystery.)

Many other damage suits dragged on beyond 1970.

Meanwhile, county and private lawyers urged the federal district court to ban new oil activity in the channel until the Interior Department first shared relevant information and conducted hearings. The government argued that the new National Environmental Policy Act requiring hearings did not apply because the leases were granted before the act took force. The court ruled against the city and county. As the case moved to the appellate court, Sun Oil lawyers charged that "these appellants simply have no private cognizable interest in federal leases on the Outer Continental Shelf." That was too much for the appellant lawyers who replied that oil companies seem to feel that the channel is "some kind of private lake to be used and exploited solely for their monetary gain." But after reviewing these and other more judicious arguments, the appeals court concluded that neither the Interior Department nor the Corps of Engineers was required to hold hearings or to share information in considering offshore oil operations.

In September, the plaintiffs asked the Supreme Court to resolve the dispute. On the last day of 1970, while the appeal was still pending, the Interior Department announced that it would—after all—hold hearings in Santa Barbara January 13-14 on applications for two more platforms.

Wally, We Hardly Knew Ye

Hickel was a defendant in many California lawsuits, but he became endeared to many Californians. He never went as far as Californians wanted him to go, but he established oil-free sanctuaries, proposed others, tightened up oil industry regulations, listened to and sympathized with anti-oil protesters. He had testified in September that miles of clean beaches are as important as oil income. He told me in the January 1970 interview that Santa Barbarans should not object to offshore oil operations on technical grounds alone; the industry engineers will always get around that, he said: "Just tell them you don't like the looks of the platforms." He was not the bureaucrat's bureaucrat. During one stuffy and confusing cabinet meeting, he whispered to a colleague, "Doesn't anyone around here ever use the word horseshit anymore?"[16]

By mid-1970 he was in trouble with the White House. He had told Nixon in May that the president was ignoring legitimate protests of the nation's youth, especially against American military intervention in Vietnam. Nixon bristled, but did not respond. But a step at a time, Hickel was shoved into the background on Interior Department issues. The president had become inaccessible. Hickel declined an invitation to a December 16-18 offshore oil symposium on the university campus at Santa Barbara, explaining that he was scheduled to be host at the White House Christmas tree lighting. On Thanksgiving Eve, it was all over; Hickel was out. GOO, which had protested his appointment the year before, now publicly protested his dismissal.

Walter Hickel supported oil and land development aggressively before he came to Washington, and he reverted to that stance when he returned to Alaska where he was again elected governor. But for one brief, shining moment, many coastal Californians felt they had a sympathetic ear at the very top of the Interior Department.

A Different Approach

For the rest of the year, the department was under an interim, acting secretary. As head of the Geological Survey, Pecora was now the dominant figure in matters of offshore oil. He did not share Hickel's attitude toward the California coast, particularly the Santa Barbara Channel and its seashore dwellers.

It was in that role that he attended the December symposium at UC Santa Barbara, and he made the most of it. In town for several days, he made several speeches outside the conference. Two dozen speakers in addition to Pecora were invited by the National Science Foundation and the UCSB Marine Science Institute, the co-hosts. But much of it revolved around Pecora.

He accused Santa Barbarans of not considering the national interest and competing demands for resources in the channel. Hickel had said Santa Barbarans "had a right to yell," but Pecora now said that "screeching and screaming aren't going to solve the problem." Pecora defended making channel oil decisions without consulting local interests. He announced that he had already approved Humble Oil's 160-square-mile deepwater oil production project in the channel.

County Supervisor George Clyde veered from his prepared text to say that people had lost faith in the word of the oil industry, the Interior Department and, most of all, the USGS which was supposed to be supervising, not promoting, the industry. He accused Pecora of consistently coming to the oil companies' conclusions. Fred Eissler of the Sierra Club castigated Pecora for an eagerness to see resources consumed rather than conserved. A panel of marine biologists bristled at Pecora's question: "What permanent damage has resulted?" referring to the channel oil spill.

Pecora said he favored public hearings on channel oil platform applications, but said there is not much Santa Barbarans can do to prevent or delay expansion of oil operations in the channel. "It looks as if the die is cast," he said.

As Pecora took the abrasive initiative and stood his ground, many at the symposium felt that Washington had declared war on Santa Barbara. Obviously Pecora was speaking for the White House as he repeatedly chastised Santa Barbarans for not supporting the president's bill to set aside part of the channel as a sanctuary, while leaving most of it open for oil production.

The response was provided by Roderick Nash, who opened the year with a declaration of environmental rights and was now closing it on the same theme.

> I reject the whole notion of compromise that the oil industry and the government hold out to us....The oil industry is not giving us anything....Let's never forget, for example, that the whole Santa Barbara Channel was a 'sanctuary' before the first rig. Those who liked it that way lose with every step toward development. Can someone or some company really *own* a piece of the environment; *buy* a mountain; *sell* a forest; *lease* a square mile of the Santa Barbara Channel?

Those were good questions for the last three decades of the century, and the federal government said the answer, alas, was yes.

Chapter Five

OPEC Changes Everything

Santa Barbarans did not have to wait long for the next crisis. In January 1971 the *Los Angeles Times* reported that the Interior Department had approved—without consulting anyone in California—adding two production platforms on the Santa Barbara seascape.[1] Three days of hearings were set for mid-month, but they would be intended only to "placate the natives," because it was a "foregone conclusion that permission to install the platforms would be granted," Interior Department officials told the newspaper.

Interior representatives explained that Platform C was needed to help reduce subsurface pressure and leakage from around Platform A where oil still flowed into the ocean two years after the blowout. Few believed the pressure-reduction theory from the start, and fewer believed in it two years later. But it gave Interior and the industry a plausible argument for more platforms. Platform C would be lined up with A and B, five and one-half miles off the county's shore, and would be owned and operated by the four companies that held the lease—Union, Texaco, Mobil, and Gulf. The other structure was Platform Henry which Sun Oil wanted to install five and one-half miles off Summerland-Carpinteria to avoid losing oil through drainage from Phillips Petroleum production in an adjacent lease.

When the Interior Department and its Geological Survey explained the purpose of the Santa Barbara hearings, they did little to dispel the notion that the decision had already been made. The hearings were not about whether the platforms should be permitted, it was explained, but only "to obtain the views of the public before evaluating the plans of the companies for installation of the platforms." Interior Department officials said that granting the leases was in effect permission to erect platforms. Any attempt to revoke that permission would lead to costly litigation, the *Times* quoted unnamed officials.

The public had different ideas, and my notes indicate one thousand persons attended the hearings, most of them to protest the installation of any new platforms. Many, of course, could not get in, and a huge press corps added to the congestion in the county administration building. City and county officials, their state and federal legislators, and citizens' groups were unanimous in their protest. The blowout was still vivid in their minds, and they had no assurance that the industry and government regulators were any more competent now than they were two years ago to manage offshore oil operations without endangering people and other natural resources.

The county was represented in Sacramento and Washington by staunch conservative Republicans: Congressman Charles Teague, State Senator Robert Lagomarsino, and Assemblyman Don MacGillivray. All testified or submitted strong statements against any new platforms. Democratic Senator Alan Cranston, of course, was equally opposed. From Sacramento came State Controller Houston Flournoy and Lieutenant Governor Ed Reinecke. Both comprised a majority on the three-member State Lands Commission. Characterizing the hearings as not much more than window-dressing, Reinecke said he regretted "that the [Interior] Department has so little regard for the exploitive use of our coastal resources and esthetics." As Republican lieutenant governor, Reinecke presumably represented the Reagan administration as well as the Lands Commission.

As the hearings ended, almost everyone agreed that they were indeed intended to placate the natives. The natives were not placated, but that phase of the protest subsided for seven months until the Interior Department issued an impact report that concluded that two more platforms "will have little effect on the marine environment." Anyway, it noted, the 1969 damage to marine life was not permanent. Again, "temporary" was the adjective chosen to describe the impact. The report was generally regarded as paving the way for approval of the two platforms.

Pressure on the President

The Santa Barbara protest machine was rolled out and refueled. Led by Get Oil Out, the environmental groups decided to appeal to President Nixon. GOO had prominent Republicans among its members in Santa Barbara, and friends in Washington. They were urged to approach the president who two years earlier had said that the 1969 blowout and its consequences had "frankly touched the conscience of the American people." Now, could the *president's* conscience be touched? If not his conscience, then maybe his political judgment. The word was passed that Nixon could lose California in the 1972 presidential election if he did not curb the offshore oil industry there.

Some concern was probably felt in the Interior Department as well. Unlike Stewart Udall's advisors who in 1968-1969 assured him that nothing could go wrong, many state and federal agencies were now cautioning Secretary of the Interior Rogers C. B. Morton of environmental risks in escalating offshore oil development. By mid-September 1971 it was learned that the decision had been moved from the Department of the Interior to the White House. It now appeared to be Nixon's decision.

But it was Morton who on September 20 announced—to everybody's surprise—that the two platform permits were denied. Morton cited "overriding environmental considerations." The platforms would be incompatible, he said, with the oil-free sanctuary proposed in the administration bill pending in Congress. The two leases involved were outside, but adjacent to, the proposed sanctuary. Whether the platform decision would be permanent would depend largely on whether Congress passed the sanctuary legislation.

In a telephone interview four days later, Morton said that the nation would not lose the oil if the platforms are delayed. "We are not wasting that oil," he said. "We need the oil, and we will need more offshore development, and we will have to figure ways to keep our energy supply up, but this oil is not going anywhere or turning sour in the ground." He said he was also influenced by the fact that "the Santa Barbara Channel is the most unique environmental situation on the entire Outer Continental Shelf of the United States." [2]

But who *did* make the decision? Two days after my interview with Morton, President Nixon was approached by Paul Veblen, executive editor of the *News-Press*, at a meeting of news executives in Portland, Oregon. Veblen asked about the decision, and Nixon replied:

> I made the decision and he [Morton] made the announcement....It's good to give the secretaries a chance—it really is. They take a lot of heat. So I told Morton, 'Rog, you go out there and make the announcement,' but I, of course, made the decision.

Morton was also at the conference, and Veblen told him what the president had said. Morton said nothing. "He just rolled his eyes heavenward," Veblen said. Owners of Platform C went to court to have the denial set aside, and owners of Platform Henry sued the federal government for more than $200 million in damages.

How Much Damage?

Elsewhere in court in 1971, Santa Barbarans won some and lost some. The Supreme Court rejected the city-county demand for public hearings before more drilling could be permitted. But a federal judge approved a $4.5 million class action award to 1,560 beachfront property owners against the owners of Platform A for damages suffered after the blowout.

Whether that adequately covered property damage is questionable, but the larger debate was over the biological harm. Dr. Dale Straughan, described by *Time* magazine as "a pretty 31-year-old zoologist," after a year's study found surprisingly little damage to marine life. The general conclusion was that the longterm damage appeared to be slight, and that "the area is recovering well." Her $240,000 study for the Allan Hancock Foundation of the University of Southern California was paid for by the Western Oil and Gas Association, the major oil industry trade group in the western states. Straughan insisted that there were no strings attached to the industry's grant. But seriously adverse findings surely could have soured the industry on future contracts with her or with USC. As it turned out, Straughan testified for oil industry projects at subsequent hearings, and got additional oil industry grants for more marine studies.

The major critic of her work was Dr. Max Blumer of the Woods Hole Oceanographic Institute in Massachusetts. He had studied the results of a 1969 oil barge spill in Buzzard's Bay, almost at the doorstep of the institute. He called Straughan's findings incomplete, her methods inadequate, and her conclusions unsupported by the evidence. In her nine hundred-page report, she contended that the volatile parts of crude oil were the most toxic, and that they evaporated very quickly, leaving many of the heavier and less-damaging elements to sink to the ocean bottom. Blumer argued that this ocean-bottom blanket of heavy crude was indeed toxic and would stay that way for a long time. Damage to coastal organisms was difficult to assess for lack of a detailed study of what was there before the spill, leaving comparisons unreliable.

Joseph H. Connell, professor of zoology at the University of California at Santa Barbara, agreed with Blumer. "It is clear that because standard scientific procedures were not employed in most cases, there is almost no basis for drawing any conclusions about the effects of oil," he said.

Straughan replied that her report clearly stated that some of the study was incomplete and that the research was continuing. She cited the lack of baseline data, and said the weather caused havoc. The spill came during heavy rainstorms and the rampaging, over-flowing coastal streams washed a deluge of fresh water, debris, and turbidity into the channel.

A Little Drilling

Time, tide, surf and sun continued their eternal cleansing, and by 1971 federal and state agencies began approving requests to resume offshore drilling. The Interior Department permitted limited exploratory drilling off Ventura County south of Santa Barbara.

The State Lands Commission approved new drilling on Standard Oil's Island Esther off Seal Beach in Orange County south of Los Angeles. True to the city's history of hostility to offshore oil ventures, the Seal Beach City Council on February 1, 1971 unanimously zoned the Pacific Ocean off its shores as municipal open space to be used for water recreation only.

Four months later, before the ordinance could face any legal challenge, the city council called it up for reconsideration. An intervening municipal election and some second thoughts had changed the mood of the council. "Despite the obvious displeasure of the clamorous audience"[3] that nearly filled the council chambers, the council voted four to one to rescind the ordinance that challenged the state's authority to allow new drilling in the "city's" tidelands.

But even state officials could not agree on how to proceed. The State Lands Commission had to rein in its staff which traditionally tended to usurp, or at least impinge upon, the authority of the commission. This time its executive officer, Frank Hortig, wanted full authority to approve drilling to deepen existing wells. Santa Barbara County Counsel George Kading and State Senator Robert Lagomarsino objected, and so did the Lands Commission members.

Hearings on tidelands drilling in August and September 1973 were contentious. The industry had Dale Straughan on its side, while the Coast Guard supported environmentalists in questioning the effectiveness of oil containment devices. From the legislature, Assemblyman MacGillivray and Senator Lagomarsino stood firm against more drilling.

Lagomarsino did not much care what the oil industry did elsewhere, but he did not want it in the channel. This was probably due to a sincere concern for channel resources, and a recognition that most voters in his district had similar concerns. He had a spotty voting record on environmental issues, but he protected the natural resources of the Santa Barbara Channel, the Channel Islands, and Los Padres National Forest in his congressional district. His bill creating an oil-free state sanctuary extending three miles seaward around the four northern Channel Islands was signed by Governor Reagan in July 1971. It was the first major channel oil legislation enacted in the state since the blowout. Nine years later, as a member of Congress, he would carry the bill to create the Channel Islands National Park.

A Frustrating Year

On the federal level there was a short flash of hope early in 1972. With considerable hoopla, Senator Henry Jackson, chairman of the Senate Interior Committee, said Outer Continental Shelf hearings were planned, not by one, but by four Senate committees sitting jointly. That sounded like extraordinary interest by the Interior, Commerce, and Public Affairs committees, in addition to the Joint Committee on Atomic Energy. They were to consider "the whole question of the leasing of Outer Continental Shelf lands." Seldom had four Senate committees met jointly on any topic.

Environmentalists concerned with the channel, however, saw some gaps. First, no specific legislation dealing with the channel was to be discussed, and no one from the channel area had been asked to testify. Industry and federal officials had been invited. Lois

Sidenberg, president of GOO and active leader in the anti-oil battle, said that the hearings were "being used to circumvent holding hearings on specific legislation having to do with protecting the Santa Barbara Channel and Atlantic coastal areas." This was especially significant since the Interior Department had indicated it would let fullscale exploration and production resume if Congress did not act in 1972 on bills to curb such activity.

Senator Jackson backed down a little. He said that GOO, the Sierra Club and the county administration could send people to testify at the hearings, now rescheduled from March 23-24 to April 11 and 18. The oil industry would be heard on the first day, and the environmentalists on April 18. Moreover, Senator Jackson indicated that the joint committee would look into how much "the risk of accident such as those near Santa Barbara and offshore Louisiana has been reduced...."

The Santa Barbara delegation crossed the continent expecting to appear before an impressive array of senators from four committees. But when the April 18 hearing opened, only four senators were on hand. And when the Santa Barbarans testified, there was only one: Senator Frank Moss, chairman of the Interior Subcommittee on Minerals, Materials, and Fuels. He had heard it all before, more than once. And he was not moved.

In the House, the reception was about the same. Wayne Aspinall, longtime crusty chair of the House Interior Committee, declined to schedule hearings on any channel oil issues. The only consolation for environmentalists was his defeat for reelection in November 1972.

By now, Interior Department agencies, particularly the Geological Survey, were not answering Santa Barbara mail nor returning phone calls. Public agencies and citizens had questions about plans for the administration's sanctuary bill, about progress toward slowing the continuing leak at Platform A, about Humble's deepwater project, and the attitude of Vincent McKelvey, the new Geological Survey director, who succeeded William Pecora, now undersecretary.

In March 1972, the city's Environmental Quality Advisory Board invited McKelvey to Santa Barbara. The *News-Press* had suggested the visit as a full opportunity to exchange views. It did not happen that way. It would be "one of our quieter-type meetings," said EQAB chairman Laurence Brundall who wanted to avoid the turmoil of Pecora's three-day visit in 1970. When McKelvey arrived on May 18 he was effectively shielded from the public. Only seven invited guests and members of EQAB were permitted to talk with him. Others could attend the meeting, but not ask questions. George Clyde, former county supervisor and a member of the invited panel, reiterated his charge that the Geological Survey and the Interior Department were "handmaidens of the oil industry." After the standard denial, McKelvey touched briefly on a few pending channel oil issues, but many more pressing matters were never raised. No one in the audience of one hundred was permitted to ask a question. But Norman Sanders, feisty UCSB geography professor, broke the rules. "We've heard all this before," he said, "and many people want to ask questions. I hate to interrupt, but as a member of the audience, I feel quite frustrated." McKelvey said he would not answer "political questions" because the USGS was not a policy-making body. He declined to talk to news reporters after the meeting; he said he had to leave immediately.

In June County Counsel George Kading tried again to obtain information about Humble Oil's deepwater drilling plans in the vast Santa Ynez unit. McKelvey replied that some information had to be kept from the public because it was proprietary, "but we note that Humble has disclosed the essential features of its development plan." That was enough for the public to know, and McKelvey suggested that anybody interested could read about it in a magazine article. Humble, too, said that there was a lot that should be kept from the people, such as "geological, reservoir and well information and other material of a confidential nature." The USGS and Humble indicated some additional information would come to light if there was a public hearing, but even then there was no explicit promise that it would all come out.

President Nixon appeared to hear the cries of California coastal dwellers from time to time, but he was not consistent in his responses. He nixed the two platforms, C and Henry, under California political pressure, but he did not worry about the political consequences of appointing—just before the 1972 presidential election—Dr. Beatrice Willard to his Council on Environmental Quality (CEQ). Willard, president of Thorne Ecological Institute of Boulder, Colorado, was known in Santa Barbara and Ventura counties for supporting a U. S. Gypsum plan to strip-mine huge expanses of Los Padres National Forest for phosphates. That plan united environmental groups into a militant coalition that fought the proposal for years in the 1960s. As environmental advisor to U. S. Gypsum, Willard urged Interior to approve the strip-mining permit. (For economic reasons, U. S. Gypsum abandoned the mining plan before turning any soil.) "Her past affiliations with industry will make it impossible for her to withstand the pressures involved in arriving at decisions and evaluations affecting the environment," Get Oil Out noted in a letter to Nixon. Willard visited Santa Barbara later to say she was a caring person without horns.

California Voters Take Charge

Without congressional and White House support, Californians looked to their legislature for coastal resource protection. For four years the legislature had debated a shoreline management plan, but it was always defeated by interests that wanted to build with a close-up view of the Pacific Ocean. This took away more and more of the seashore from residents who did not own oceanfront property. Californians did not want 1,100 miles of Waikiki Beach.

Philip Fradkin of the *Los Angeles Times* wrote:

> California, with one of the most dramatic and lengthy coastlines in the nation, is almost alone among coastal states with no legislation on the books to protect this resource. For the fourth year in a row, coastline legislation has been introduced in Sacramento. But the outlook for passage of any bill this year is bleak and all strong measures are virtually dead. So conservationists are attempting to get a strong initiative measure qualified by today [June 19] for the November ballot.[4]

Their patience exhausted, environmental leaders had drafted a coastal protection initiative based on four years of debate in legislative committees. They knew the objections, the desires, the strengths and weaknesses of coastal resource plans.They knew where to compromise, where to stand fast.

In November, fifty-five percent of the voters approved a coastal initiative that would create a State Coastal Commission and six regional commissions that would in turn set criteria for coastal development. Each waterfront city and county drafted its own coastal plan under state standards. Until the plans were approved by the new State Coastal Commission, all coastal permits would be reviewed by regional coastal commissions.

This meant not only that waterfront construction would be limited and subjected to higher standards, but that tidelands activity would come under California Coastal Commission review. Even beyond that, under the 1972 federal Coastal Zone Management Act (CZMA), every state with federally-approved coastal plans won the right to "object" to Outer Continental Shelf projects that would conflict with the state's coastal plan. Such an "objection" would stop a project—such as offshore drilling—unless overruled by the secretary of commerce who administered the CZMA. That meant that the Commerce Department could be pitted against the Interior Department, adding a layer of review of OCS oil operations. Largely because of legal challenges by the oil industry, however, the California coastal plan did not win federal approval for several years.

Exxon Audacious, No Longer Humble

After a long search for a new name, Humble Oil became Exxon on January 1, 1973. Exxon officials said they wanted a name that was not derogatory or embarrassing in any language. But some environmental opponents soon identified Exxon as "the sign of the double cross." In any case, Exxon was no longer Humble nor humble. Together with its bidding partners, Exxon had in 1968 won leases that spread over more than 160 square miles in the western end of the Santa Barbara Channel. The companies now submitted an audacious plan for producing in water more than eight hundred feet deep from five platforms standing as high as ninety-story buildings. None of the thirteen platforms then in the Santa Barbara Channel were in more than two hundred feet of water. Exxon was in for the deep plunge, and most Santa Barbarans saw it as a perilous experiment. There would also be an enormous processing plant on thirty acres onshore, and a mooring for loading tankers.

But the Interior Department's 1,278-page impact report maintained there was no reason to worry. Technology had advanced superbly, and new regulations assured safe operations and strict enforcement. The caprock was thousands of feet thick, not flimsy as in Union's Dos Quadras field. Exxon would have oil spill containment gear at the ready. Federal safety standards were now as high as those of the state, which had frequently boasted that the 1969 blowout would not have happened under state control.

The Interior Department scheduled a two-day October 1973 hearing on the project. After Exxon's witnesses consumed the first three hours, the department's hearing examiner said everyone else would get ten minutes each—except the California Chamber of Commerce which would get a half hour. This did nothing to placate the natives. For two days Exxon's safety record was attacked, and the platforms were described as navigational hazards, a threat to marine life, and a visual blight. Exxon, of course, had friends among the ninety-one who testified. Residents of Santa Barbara's "north county," where oil is a major part of the economy, supported the project, as did several oil-related companies. The hearing panel of thirteen Interior Department officials took this testimony back to Washington to help them decide whether the impact report needed changes. But little of the testimony dealt with the validity of the report; most witnesses talked about the merits of the project, not of the report. Such was the nature of oil hearings in Santa Barbara for a generation. The specific issue was generally ignored. Witnesses came to say they did or did not like offshore oil ventures.

The Energy Crisis

The war in Vietnam helped activate the first major federal venture into oil production in the channel in 1968, and another foreign war changed everything there five years later. Late in 1973 events in the Middle East delivered a jolt that was felt in industry, oil-heated living rooms, and especially in car-crazy California. Israel was attacked by a coalition of Arab nations, and the American government felt compelled to supply Israel with military hardware. On October 20 Saudi Arabia, the Middle East's greatest oil producer and traditional friend to Washington, reduced oil shipments to the United States and boosted the price. Other Arab members of the Organization of Petroleum Exporting Countries (OPEC) joined the anti-American campaign. At that time the United States was importing thirty-six percent of its oil, and OPEC nations were providing forty-eight percent of American imports. Not all OPEC nations joined the effort, but enough to create the great American "energy crisis of 1973."

This changed attitudes in the White House and in Sacramento. When the Senate Interior Subcommittee on Minerals, Materials, and Fuels met in Washington the administration reversed itself and withdrew its plan for oil-free sanctuaries in the Santa Barbara Channel. "We have a very difficult backdrop for a Santa Barbara decision today from the one we had in 1969 or even earlier this year before war in the Middle East further disrupted our patterns of energy supply," the subcommittee was told by Stephen Wakefield, an assistant secretary at Interior. He said the embargo had made an already serious situation worse.[5] Three days before the embargo was imposed, Congressman Edward Hebert of Louisiana, chair of the House Armed Services Investigating Subcommittee, indicated he opposed dipping into the Elk Hills petroleum reserve at that time. The Nixon plan was to use oil from that reserve to compensate for what would be lost in the Santa Barbara Channel sanctuaries. If Hebert would not go for it, the chances were slim in the House.

Despite this turn of events, Congresswoman Patsy Mink brought her House Interior Subcommittee on Mines and Mining on November 17 to hear more local views on offshore oil. Congressman Teague, who had carried the administration's bill in the House, let it be known he felt betrayed. He said he would introduce legislation requiring congressional approval for any new oil production in the channel, subjecting the Interior Department to congressional permission. Bill Gesner, former offshore drilling crewman, reiterated what he said was firsthand knowledge of sloppy drilling operations. And he cited a recent Government Accounting Office report to substantiate his accusations. The Santa Barbara protest was recited in full once more by GOO, Sierra Club, League of Women Voters, the Audubon Society, the Carpinteria Valley Association, and the California Coastal Alliance. For her part, Mink said at the end that "I feel generally aligned with those who testified in favor of an energy reserve" instead of immediate exploitation in the channel.

For all the good it did. The "energy crisis" had momentum now that would not be slowed by appeals for more deliberation. The appeals now were for increasing electric and petroleum production to meet American demand. For the next couple years energy corporation executives, government officials, and others who should have known better, insisted that energy output must be increased three to seven percent per year for the indefinite future. Caught up in the fear and ferver of the time, audiences nodded in agreement, but anyone with a pencil and paper and a sixth grade education could have easily determined that this kind of escalation was physically impossible beyond an extremely limited horizon. Nor, as it turned out, was it necessary.

But the campaign snowballed, and over the usual protests of the usual Santa Barbara delegation, the State Lands Commission on December 11, 1973 lifted its five-year drilling ban in California waters. For a start, it approved drilling on the eight operating platforms in state waters. Exploratory drilling and new platforms would be considered later on a case-by-case basis. Commission Chairman Houston Fluornoy said the decision had nothing to do with the "energy crisis," but the commission staff and Interior Department witnesses emphasized the need for more petroleum sources in light of sudden shortages.

American oil production began dropping off in the 1970s, made a brief recovery in the mid-1980s, and has steadily declined since. As consumption remained high, imports mounted. Writing early in 1974, Stewart Udall (and two co-authors) noted that the United States was still "making extravagant consumption plans for the future and operating on the illusion that we are still the world's petroleum powerhouse when, in fact, we were slowly becoming a have-not nation....[B]ig oil-men, who clung to the old myths that served them in the past, were continuing to prate that if we let them make enough money they would find the oil and eliminate the shortages."[6]

The nation's energy policy was still described, if it could be identified at all, in two words: *laissez faire*. Three words, if you add gluttony.

"Accelerated Development"

OPEC's embargo was not airtight and did not last long, but Washington understood at least some of the ramifications, and it panicked. The embargo dramatized the vulnerability of the nation's energy supply. Policy changes were required. Options included giving in to OPEC, rationing oil products in the short run, ending oil squalor and embracing energy efficiency, and increasing domestic oil production. Nixon, fully recovered from the personal grief expressed over the 1969 blowouts, selected the latter.

In January 1974, two months before the embargo was lifted, the president instructed the Interior secretary to offer ten million acres of ocean annually to the oil industry, starting in 1975. "If successful, this sudden acceleration of OCS leasing would double in a single year the total acreage leased during the previous twenty-one years of federal OCS leasing, and would open up hitherto untapped areas to exploration," a congressional study warned.[7]

Southern California was the first big target. The Interior Department had already decided to offer 6.1 million acres extending from the Channel Islands to the Mexican border, and under the acceleration another 1.6 million acres were added to Lease Sale 35. There was little chance that all 7.7 million acres would ultimately be auctioned. The Interior Department would whittle it down after the oil industry picked it over and told the department what parts it wanted.

The sale would be south of Santa Barbara, this time off Ventura, Los Angeles, Orange and San Diego counties. A few oil islands and platforms were there already from earlier state tidelands leasing. But in these coastal counties the ramifications of the federal proposal were not fully comprehended for several months.

At a state Assembly committee hearing in Santa Monica in April 1974, a few local officials from coastal cities protested. The Manhattan Beach City Council, for instance, stated it had had all the offshore oil experience it wanted. Officials from Santa Monica and Los Angeles also told Interior officials to look elsewhere for oil.

But by July 21 when an Interior Department panel came to Santa Monica to hear local comment, hostility had gathered momentum. What was once mostly a Santa Barbara protest had now become a Southern California demonstration of outrage. Watching the confrontation between private citizens and state and local officials on the one hand, and the Interior and oil industry officials on the other was like seeing a new production of an old movie, this time on a wide screen, with a bigger cast, and with the sound turned up.

Even before the July session opened in Santa Monica, the approach to the civic auditorium was lined with booths and placards erected by irate groups, many of which conducted well-attended news conferences. Inside, protests were heard from the city councils of Manhattan Beach, Redondo Beach, Hermosa Beach, Torrance, Palos Verdes and Santa Monica. Mayor Tom Bradley voiced Los

Angeles' opposition, and the Malibu Township Council joined the chorus.

The *Oil and Gas Journal*, a leading trade publication of the industry,[8] noted that ". . .the general attitude of many officials in cities along the beaches was that the area should not be opened to drilling." It was about the mildest way it could have been expressed.

A Broader Dispute

Polarization over oil development probably was never so complete nor on such a broad scale as in the controversy over the federal government's 1974 plan to accelerate leasing off California. Santa Barbara's seven-year struggle was longer, but compared with the number of people involved and the boldness of the contending forces, it was only preliminary to the new and larger battle.

This was dramatized in September in Santa Monica before the Senate Commerce Committee's National Ocean Policy Study group. The dispute included a confrontation between Senator John Tunney of California and Senator Ted Stevens of Alaska. Tunney repeatedly asserted that the Interior Department was rushing into the vast leasing program without seriously consulting state and local officials. He said it jeopardized the California Coastal Plan before it could be completed. Stevens asked why California should not produce its share of oil from offshore. Alaskans did not especially love offshore oil production, he said, but they were preparing for it.

Lined up against the leasing plans were mayors and city councils of several Southern California communities, Senators Tunney and Cranston, a majority of state assemblymen, a coalition of environmental groups, more than 200,000 petition-signers, the State Coastal Commission, and the state attorney general's office.

Familiar Debates

Arguments for and against the leasing program had become familiar, and were repeated at each hearing throughout the 1970s and beyond.

Industry and Interior officials said the nation must have an expanded and reliable domestic source of oil, and offshore seemed to be the greatest source still only lightly exploited. In addition to oil, on which the nation's economy and security depended, the effort would provide jobs and public revenue. Indeed, the vaulted

American standard of living depended on a steady, reliable domestic oil supply. And then some. To maintain this standard would require a three to six percent annual increase in available oil.

There had been mishaps in the early years of offshore oil experience, the arguments continued, but technological improvements, public regulation, and better trained workers now reduced the risk of accidents to a small and acceptable level. Consequently, those who raise environmental objections were often merely obstructionists or suffered from a NIMBY (Not In My Back Yard) complex. One must consider the national interest and abandon provincial thinking. And people (who spend money and vote) had to come before the interests of lesser species.

The fact that arguments of the industry and the Interior department were almost identical made it easy for opponents to accuse the two of being in bed together. Sometimes in off-guarded moments federal officials acknowledged this by, for instance, referring to the industry as "we," and their indiscrete comments had to be quickly denied or "clarified."

Checks and balances were weak. The Department of the Interior had the conflicting duties of promoting oil production and policing the oil industry. It was trying to do both when it waived a vital regulation on Platform A in 1969.

On the other side were citizens and many state and even federal officials who charged that the Interior Department sold offshore leases without knowing enough about what they were worth. The federal government did no exploratory drilling, and very little geophysical work. Only the oil companies knew much about the economic potential of the lease areas, and they shared no data with their competitors and very little with the federal government. Consequently, the Interior Department did not know how much to expect or require in bids from lease to lease. Federal revenues suffered.

The Department of the Interior often knew little about the biological resources of lease areas, and would therefore be unable to accurately assess the damage in the event of a spill. If one does not know what was there *before* contamination, it is hard to compare it with conditions *after*. Sometimes environmental studies came *after* leases were sold. It was too late then to cancel a lease, and difficult to impose environmental restrictions that conflicted with lease terms.

Coastal communities protested that the federal government withheld information that would help in evaluating leasing programs. State and local officials complained they were more often ignored than consulted before federal decisions were made. They lacked a voice in the management of their resources.

Opponents challenged the offshore industry's ability to do its job safely and cleanly, and said that even when everything went right the industry was a coastal blight. The mere presence of platforms, tankers, and the accompanying onshore industrial complexes intruded on habitats and crowded out other uses of coastal resources. This included commercial and sports fisheries, and waterfront recreation.

The concern was not only at the state and local level. Several federal agencies were disturbed. The President's Council on Environmental Quality wanted, among other things, stricter standards for discharging drilling muds, drill cuttings and other waste into the ocean, and said that all new tankers in coastal trade should be built with double bottoms, and with ballast tanks segregated from cargo tanks. It urged stricter inspection and enforcement of offshore oil regulations, and a greater voice by the states in coastal resource management.

A review committee of the National Academy of Sciences supported the CEQ recommendations and added a few of its own. It maintained that Interior must, by its own efforts or through corporate contracts, find out in advance how much oil is available in the leases they award. The National Ocean Policy Study, a Senate Commerce Committee research effort, drafted probably the most complete list of complaints about the leasing process at that time. Its reports included the gamut of opposition testimony heard at hearings in the 1970s.

Even if ambitious offshore leasing could eventually be justified, critics said, it was premature in the 1970s. The nation lacked an energy policy that would indicate a need for much new offshore oil production. The federal government relied on the free market to determine an "energy policy." And the "free market" was defined by the oil industry.

Several state and federal agency officials said that the oil industry had neither the financial resources, personnel, information, nor equipment to cope with a ten million acre annual leasing program. The result would be anemic bidding on each sale as the industry tried to spread its resources across the vast lease areas. Federal revenue would suffer as bids per lease would be far below their real value.

A New Administration

The Ford administration, in office since the August 1974 resignation of Nixon, backed away from the ten million acre a year plan and now proposed leasing no more than seven million acres annually from 1975 through 1978. Even that, however, was audacious, and California and several East Coast states informed the new president they would fight the plan.

Neither side was giving much ground, and Interior kept California coastal communities on edge with each new leasing announcement. In October, Interior moved the Southern California leasing area closer to shore as it withdrew waters farther offshore which were in or near the Navy's Pacific Missile Range. In mid-November the Interior Department disclosed plans for leasing north of Point Conception along Santa Barbara's west-facing coast and possibly all the way to San Francisco. San Luis Obispo County, previously undisturbed by the leasing battles, suddenly felt ambushed. The following headline ran across the top of the page of the November 14 *San Luis Obispo Telegram-Tribune*, almost begging for an exclamation point.

> U. S. wants oil drilling
> off SLO County's coast

Pressing on, the Interior Department scheduled February hearings in California, the East Coast and Alaska to brief the public on its extensive leasing plans and hear public response. The department knew the California attitude, but even then must have been surprised. Three coastal communities had officially protested at hearings in April 1973. The number grew to eight by July. Before the Interior entourage took its seats for the February 1975 session in Beverly Hills, forty-one coastal cities and counties had their signatures on a five hundred-page protest. Almost as imposing as its report was the group's name: The Council of Local Governments Concerned with Federal Proposals for Offshore Oil Development on the Outer Continental Shelf.

Its arguments, already noted, were not new, but the rapidly-escalating breadth of the rebellion would have been regarded by almost any other agency as overwhelming. Not by the Department of the Interior .

(On the East Coast, the reception was not much more cordial. In April 1975, the six-member Mid-Atlantic Governors Conference called on Congress to prohibit offshore leasing before the states got federal environmental and economic reports, and demanded a greater voice in the leasing process and some revenue from any ensuing production.)

"Vocal, Fanatical Minority"

An Interior Department delegation returned to California in August 1975 for three days of hearings on leasing off Southern California. More than one hundred persons signed up to testify, most of them in opposition. Supporting the industry was a raucous delegation from the Long Beach Chamber of Commerce. Otherwise, the testimony from outside the oil industry opposed leasing.

Philip Verleger, a Los Angeles lawyer who spoke for two hours on behalf of the oil industry, called the opposition "a vocal and fanatical minority." Statistics indicated otherwise. Three months earlier, Exxon was threatened with a referendum that could have prevented the construction of a massive processing plant in Santa Barbara County. Exxon and its friends spent about $300,000 (twelve times the opposition's budget) to defeat the initiative by a one percent margin.

Also part of the "vocal and fanatical minority" were seven county agencies, the city of Santa Barbara, five non-government specialists and three county consultants who produced a report of about two hundred pages of opposition to federal leasing plans.

Albert Reynolds, the county's environmental quality coordinator, stated:

> In a nutshell, we found that [the federal] Draft Environmental Statement is grossly inadequate in its failure to display and document the full environmental impacts associated with the proposed gas and oil development in the OCS lands of the Santa Barbara Channel.

The Board of Supervisors agreed, and approved throwing the two-pound book at the Geological Survey panel. "Let me say that this report represents community action at its best," Reynolds told the federal hearing officers. It was not the raving of a "vocal and fanatical minority;" it was the work of specialists in geology, law, air quality, meteorology, land use planning, environmental management and other fields, all of whom found the federal report lacking.

The Interior Department panel took all the testimony back to Washington to prepare for a Southern California lease sale before the end of 1975. California's new young governor, Jerry Brown, did not wait for the details. His Office of Planning and Research sent notice that the governor did not like the procedure.

Governor Brown said the leasing rules did not require enough information from the oil companies, did not provide sufficient authority to stop hazardous operations, and flouted federal legislation requiring hearings and impact reports. The California Coastal Commission staff concurred, charging the Interior Department with a "lease now, plan later" approach.

Undeterred, Interior officials prepared the Secretarial Issues Document (SID) that would help the secretary decide whether to go ahead with a lease sale, and under what conditions. Los Angeles and state officials asked to see the document. They were told it was for the secretary only, and would not be made public. That procedure was to become a major center of dispute in future lease sales.

Thomas Kleppe, recently appointed secretary of the interior, eased some tension by deleting Santa Monica Bay from the sale. But in San Diego the big crisis headline type was called up on page one of the October 24 *San Diego Union*:

> OIL HUNT DISCLOSED
> 94 MILES OFF S. D.

Other headlines that day in the Union: "Officials Here Voice Dismay At Oil Drilling; Several Express Outrage On Being Kept In the Dark For Two Months." "GOVERNMENT AGENCIES IN CONFLICT; Offshore Oil Drilling Snarled By Bickering."

After all that expression of panic, the sale was an anti-climax. On December 12, 1975, in Los Angeles, the industry submitted high bids totalling slightly more than $438 million, far short of expectations. The Bureau of Land Management, which conducted the sale for the Interior Department, had hoped for—and expected—one or two billion dollars. Only 70 of the 232 tracts offered received bids. The companies bid mostly for leases off San Pedro in the Los Angeles-Long Beach area, where wells were already producing from offshore extensions of onshore fields. The closest bidding to San Diego was 125 miles at sea.

The experience should have led to a serious review of how to conduct leasing in the future. The Interior Department had caused panic along the coast by offering, at the outset, millions of acres offshore to the oil industry. Coastal communities from San Luis Obispo to San Diego suddenly faced planning for an unprecedented and unexpected industrial onslaught for which they were totally unprepared. The fact that the sale fell on its face did not take away from the disruptive threat that was posed in the meantime. There was no need for such a threat. The Interior Department could have selected small potentially-productive areas for sale. This of course would have required conducting its own research, contracting the work out to private corporations, or purchasing the information from people who had it.

Instead the Interior Department, which admitted it was "flying blind" in offering leases, simply blocked out enormous areas for lease sales and waited to see what would happen. It was another manifestation of letting the "market" determine energy policy. But it wasted staff time in Washington and in coastal communities; it challenged the oil industry to plan for over-extended expanses of ocean; and, it deprived Interior of information needed to determine whether bids were adequate.

But Washington did not slow preparations for more West Coast lease sales. In fact, the Interior Department had broadened its Pacific horizons to include everything over which it had authority—from Mexico to Canada. Lease sale 48 would cover much of the same area as the disappointing 1975 sale, plus possibly the Santa Barbara Channel. Lease sale 53 would run from Point Conception in Santa Barbara County all the way to the Canadian border. What part of this enormous stretch of the Pacific would eventually be leased would depend, again, on what the oil industry was most eager to exploit. Some environmental, military and other concerns would also be considered.

The two sales would occur in separate regions. But with Point Conception as the dividing line, Santa Barbara County would be involved in both. South of Point Conception is the Santa Barbara Channel. To the north is the county's west-facing coastline. No other county would have to prepare for both sales.

In October 1975 Kleppe said that the Santa Barbara Channel and Santa Monica Bay probably would be excluded from sale 48, but before the end of the year he gave in to the Bureau of Land Management and the Geological Survey and indicated that both areas probably *would* be part of the sale. (The BLM at that time prepared and conducted the sales, and the USGS applied and enforced regulations after the sales.) As usual, industry was eager to buy more of the channel. It also displayed unusual interest in the Santa Maria Basin off Santa Barbara's west-facing coast. Much had changed since the first sale there in 1963 when Santa Maria crude was selling for one dollar a barrel. Oil was discovered after that sale, but at that price no company bothered to produce it. Now, with domestic oil averaging $8.19 a barrel, it was worth another look.

Lease sale 48 was originally set for 1977, and sale 53 for 1978. Things always take longer than planned, and the sales would be delayed until 1979 and 1981, respectively.

An Energy Policy?

In mid-1973 oil prices rose and supplies became tight. Even before the October OPEC embargo, the plight was described as a crisis. In an article that started on page one and occupied much of three inside pages, the *Los Angeles Times* attempted to explain the shortage.[9] The report attributed the shortfall to a lack of planning for increased oil production. It neither laid blame on waste, nor suggested that conservation, alternative fuels, transportation alternatives, and energy-efficiency might help alleviate the "shortage." More oil was all that mattered.

Frequently when testifying at oil hearings, critics of the more-oil policy were asked if they walked to the meeting—or maybe rode a horse. They had, of course, arrived in gas-guzzling, energy-inefficient, air-polluting autos. That is all industry had offered. The auto, highway, rubber and oil industries—in league with government at many levels—had designed a system that relied heavily on the automobile. Mass transit and other more energy-efficient ways of getting about were undermined in favor of systems that required oceans of petroleum.

Among public agencies, energy efficiency and conservation and alternative fuel sources were still considered quaint—even noble—objectives but lacking any practical potential. An ever-increasing oil supply was all that would save the American economy and standard of living.

But the people were ahead of their leaders. In their private role, consumers began to reverse a trend that had been regarded as normal, inevitable, and even healthy. After a decade of increasing oil consumption, ranging from 4 to 5.7 percent annually, Americans in 1974 burned 2.3 percent *less* gasoline than they had in 1973. That stemmed from higher gasoline prices and a recession, not just from a conservation ethic, but it demonstrated that a large increase each year was neither inevitable nor necessary. And it was done with virtually no political or industrial leadership.

The falloff in demand for all petroleum products in 1974 was even greater than the decline in gasoline use. American demand for all petroleum products was down 3.3 percent, an impressive trend when one considers the great annual increases during the preceding decade. But the *Oil and Gas Journal* was not impressed. "At best," it editorialized, "conservation can only reduce the rate of growth in energy demand....It is highly doubtful that conservation alone can lower consumption to an acceptable level without stagnating the economy or drastically altering a nation's style of life."[10] The major oil companies professed to support fuel conservation and efficiency, but responded grimly to any decline in oil consumption.

The *Oil and Gas Journal* editorial concluded that conservation "should be viewed as only one of the vital efforts needed—along with strenuous action to enlarge the supply base—in assuring a nation ample future energy."

And when the oil is gone?

While the public was waiting for political leaders to catch up, voluntary conservation achievements were being made without political leadership. Environmentalists were recommending:

- A major governmental effort to encourage the use of alternative energy sources and to increase energy efficiency.
- Governmental guidance and example in energy conservation.
- A reverse rate structure for energy users, putting the higher unit rate on big users, not on small consumers.
- Legislation lifting discrimination against recycled products.
- A stiff tax on high-powered, heavy autos.
- A stringent miles-per-gallon requirement for all new cars.
- Policies fostering mass transit.
- Communications alternatives to transportation.

A Cumbersome Bureaucracy

But even if public policy had shifted to "soft energy paths"—which it had not—a strong petroleum production program would have been necessary in the meantime. That posed a problem on the Southern California coast and particularly along the Santa Barbara Channel. The area was impacted by more than nearby offshore fossil fuel production. Oil from Alaska's North Slope and natural gas from Indonesia, for instance, would converge on the Santa Barbara Channel.

Public policy and bureaucratic jurisdictions were not devised to cope with this. The inability to consider all these developments in a package—each in relationship to the others—stemmed from the fragmentation of planning and regulatory authority among scores of agencies.

Not only was it impossible to adequately relate one project to another, but often even to consider a single project as a package. For instance, the Exxon production proposal started at a platform in federal waters, continued via pipeline through federal and state waters, onto a county beach, through the Coastal Commission permit zone, and to a processing plant where the county had authority, back out through the coastal zone with another pipeline, through the state intertidal zone again and into state waters to a tanker buoy that required the approval of at least two state agencies.

From there tankers would take the oil through state, territorial, and international waters before leaving the channel. The Coast Guard had established sealanes through the channel for tankers, but they were only recommended courses that could not be enforced. The Army Corps of Engineers, meanwhile, decided where drilling rigs and production platforms should be in relationship to the shipping lanes.

Several other state and federal agencies would enter later to set and enforce standards for air quality, ocean dumping, supply boat routes, natural gas transportation, and other offshore industrial activities. No one agency had authority to co-ordinate all the components of this single project, much less look at a combination of such overwhelming endeavors.

That was the bureaucratic complexity for just one company's project. In the mid-1970s at least three oil companies wanted to expand coastal plants or build new ones on the channel coast.

Efforts at coordinating the polychotomous tangle failed, as no public agency wanted to dilute its authority in another's domain. It was in this morass that citizens sought to protect natural resources from the devastating impact inflicted by the exploitation of one natural resource—oil. Often environmentalists used the unwieldy system to delay or modify an unwanted development. But the oil industry had unmatched funding for the lawyers, consultants, permit fees, public relations and campaign contributions to get their plans approved. Approval took time, of course, and plans were modified in the process. Both sides used the system with skill and/or financial resources. But in the absence of a forward-looking, enforceable national energy policy, the system served the interests of the oil industry, not the public.

With more money than talented management, the industry sought to enhance its image. Its efforts were often extensive and expensive, but not overwhelmingly successful. In February 1976 four major professional petroleum associations invited fourteen California journalists to a three-day forum at Asilomar on Monterey Bay. They were to be enlightened by eighteen oil executives from the American Association of Petroleum Geologists, the American Association of Petroleum Landmen, the Society of Exploration Geophysicists, and the Society of Petroleum Engineers.

As represented by these instructors, the industry was not eager to reduce energy consumption through efficiency, conservation, or other means. The answer to demand, they insisted, was greater production. If fossil fuels were not enough, nuclear or thermal energy would help. A continuing increase in the rate of energy consumption was assumed. But it was not called consumption. It was referred to as "need," " demand," and "appetite." There was no discussion of how much of the need, demand, and appetite was waste encouraged by the industry.

During the same month, two government-sponsored reports concluded that Americans were wasting half the energy they consumed.[11] But the oil industry representatives insisted that Americans must continue consuming at the then current rate—and then some—to maintain their freedom. The nation's freedom and strength, they said, come from high energy consumption. And the nation has sufficient energy resources to continue an escalating rate of energy consumption if only private enterprise would be unshackled and permitted to produce these resources, the journalists were told.

One speaker referred to moratoria on drilling in the Santa Barbara Channel as a form of political obstruction to energy development. But he declined to elaborate, and another faculty member objected to his even being asked to discuss it further.

In private discussions, some faculty members appealed to national chauvanism. One felt that "Ay'rabs" had "stolen everything they got" in oil technology. He and others denounced the OPEC nations for setting the price on their own oil. And two or more expressed distress that the Soviet Union and China were developing their oil fields with considerable success.

One said he was sure Venezuela would no longer discover new oil fields because, under nationalization, who would do the exploring? No one thought to ask who was exploring in China and the Soviet Union whose petroleum progress was causing such trepidation. During the formal sessions, solar energy was dismissed as only fifteen percent efficient. But even fifteen percent efficiency from an easily converted, perpetual source (which was already providing most of our energy) was immensely preferable to a high-cost energy source that is limited, polluting and at best thirty to sixty percent efficient.

"For the next ten to twenty years," one speaker said, "you have no alternative than to bet with us that we're going to find that oil." But Americans and the rest of the human race were likely to persist beyond twenty years, and almost nobody in Washington and the industry was thinking much about that.

Chapter Six

The Oil Industry Regains Momentum

In July 1974, Union Oil and its three partners agreed to pay $9,465,000 to Santa Barbara city and county, the city of Carpinteria, and the state for damages suffered in the initial 1969 blowout. After four and one-half years of persistent legal effort, the victors were exuberant. "It's the biggest oil pollution damage payment so far—bigger than the *Torrey Canyon* payment," said James Christiansen, Carpinteria city attorney.

"It isn't often that you can settle lawsuits at your own figures," said Barry Cappello, Santa Barbara city attorney.

And Marvin Levine, deputy county counsel: "This is going to ring bells in Washington." He said the amount of the settlement would motivate the Geological Survey to regulate the offshore oil industry more rigorously, and added that "oil companies, too, are going to be more cautious." Perhaps, but they were not losing much ground at the moment.

During the final days of Governor Reagan's administration, the lame duck State Lands Commission approved drilling thirty-six new wells on four Standard Oil platforms off Summerland and Carpinteria, and seventeen new wells from the Atlantic Richfield platform off Coal Oil Point. These were the first state drilling permits approved since the 1969 blowouts, and probably posed little environmental risk. But they were resolutely opposed by Santa Barbara city and county officials, the area's state legislators, and environmentalists. Perhaps most galling was the fact that the commission's mandate evaporated in the November election. The incumbent commissioners would be out of office in less than two months.

California Lands Commission members were not elected as such, but two were members ex-officio as lieutenant governor and state controller, which are elective offices. The third member was the state finance director who was appointed by the governor. The retiring governor was Ronald Reagan, and the governor-elect was

Edmund G. (Jerry) Brown Jr., no friend of the oil industry (nor of Ronald Reagan). Mervyn Dymally was elected lieutenant governor, and Ken Cory state controller, both close to Brown on offshore oil issues. The ideological turnover was complete.

During its first month in office, the new commission rescinded the Standard Oil permits and suspended ARCO's permits pending more environmental and technical review. In October 1976 the new State Lands Commission approved the Standard Oil drilling, but the company still needed approval from the California Coastal Commission. In ARCO's case, the California Supreme Court ruled that the company had a vested right to resume drilling based on extensive platform drilling before the California Coastal Act came into play.

In federal waters, meanwhile, the drilling moratorium was virtually dead, and by late 1976 exploratory drilling vessels were again churning the Santa Barbara Channel and now even south of the channel where leases had just been awarded. And in anticipation of future lease sales, at least three geophysical survey vessels were testing the formations under the ocean bottom south of the Channel Islands.

Embattled Exxon

During the first half of the 1970s, for obvious reasons, Union Oil was the bad boy on the block. By 1975, however, Exxon had become the chief source of contention. In February, the Santa Barbara County Board of Supervisors, on a three to two vote, cleared the way for Exxon to build a large processing plant in a coastal canyon twenty miles west of Santa Barbara. Robert Kallman, who later became part of the Minerals Management Service[1] team in Washington, had replaced George Clyde on the Board of Supervisors, tipping the balance in favor of oil development. (At times, however, Exxon tested Kallman's patience beyond his limits.)

In an attempt to reverse the supervisors' action, GOO organized the County Environmental Alliance in order to place the issue on the ballot. Exxon had been through this kind of struggle seven years earlier, when the company was still called Humble. In a county referendum then, the voters aborted company plans to build a processing plant at Carpinteria.

The 1975 alliance quickly raised more than twice the number of signatures needed to put the issue on the ballot. This time the company took no chances. Before the May referendum, Exxon and its friends poured nearly $300,000 into the campaign, more than twelve times the $23,793 spent by the Alliance. On the South Coast where the plant would be built, the vote was fifty-nine percent against it. But the north county, where oil was not considered a monster, the pro-Exxon vote saved the project—by a one percent margin.

The dispute did not end there; indeed, it became more contentious. The plant near the ocean required a California Coastal Commission permit. The battle went through a series of heated hearings until March 1976 when the commission voted nine to two to approve the project. But largely because of one requirement, Exxon refused to accept the permit. The Coastal Commission noted that Exxon could tanker its oil for five years while the state studied the feasibility of pipeline transport. The commission would then decide whether to require the company to abandon tankering in favor of pipelines. Exxon did not object to the study, but insisted that the company—not the Coastal Commission—should make the choice between the two systems. But the commission insisted that the decision was a public responsibility.

Just as the final hearing opened at the Airport Park Hotel in Inglewood, the commission received a telegram from Kent Frizzell, Undersecretary of the Department of the Interior, warning that if the Coastal Commission did not approve a tanker terminal, Exxon could abandon the onshore project and install a floating storage and processing plant—a converted oil tanker—near Platform Hondo beyond local and state control. Tankers would then take on oil from the offshore facility. This was not new information offered in a spirit of cooperation. Exxon had earlier made this "threat," and Frizzell's telegram, too, was perceived as a threat.

Joseph Bodovitz, the Coastal Commission's executive director, called Frizzell's telegram "outrageous and highhanded" pressure on the commission, coming with virtually no consultation with the commission staff and with no study of alternatives to the floating processing plant.

With the Interior Department's help, Exxon managed to confuse the issue so badly that some of the press left the hearing with the impression that a permit had been denied. United Press International reported as much. The *Los Angeles Times* reported the permit was "in effect rejected." By the time the headline-writer got through with it, readers were informed that "Coast Panel Rejects Exxon Onshore Unit."

The protracted struggle continued before county and federal agencies, and Exxon carried its cross through it all. As late as October 1977, Phillip Verleger, Exxon's legal spokesman, three times told an Environmental Protection Agency panel that the California Coastal Commission had denied permission for the onshore plant.[2]

Whether it would process its oil onshore or offshore, Exxon would in any case produce it from Platform Hondo in water 850 feet deep in the western end of the Santa Barbara Channel. The platform was so big it came in two sections on barges from an Oakland shipyard. The sections were welded together at sea off Santa Barbara. Being buoyant, it was towed in floating horizontally. As the lower legs were flooded, the 865-foot 19,000-ton steel framework began tilting bottom-first toward the ocean floor. Before hundreds of news media and other spectators at sea and in the air, the platform frame on June 23, 1976, settled perfectly into position. When the working decks were added, it stood 950 feet high, 95 feet of which was above water. It was then the world's tallest offshore oil platform.

The following month Union Oil's Platform C, after languishing prostrate onshore for seven years, was approved for installation in the channel. It was built in 1969, but the Interior Department revoked its approval after the Platform A disaster. Now the department felt confident to let it join eight other platforms standing in line along the Rincon Trend off Carpinteria, Summerland, Montecito, and Santa Barbara.

A Changing Cast

By the mid-1970s, then, the offshore oil industry enjoyed a new momentum as the moratoriums of the post-blowout period expired and new industrial activities were approved. But the future of the industry was far from certain. A new cast of political leaders had emerged. Congressman Charles Teague, who had carried Nixon's moratorium legislation even longer than Nixon did, died on New Year's Day 1974. State Senator Robert Lagomarsino succeeded

Teague, his political twin, in Congress, and Omer Rains, a Ventura County Democrat, replaced Lagomarsino in the state Senate. Before the year was over, Nixon resigned, following shortly the resignation of his vice president, Spiro Agnew. Gerald Ford, who had been elected neither president nor vice president, suddenly found himself president of the United States. He continued Nixon's policy of "accelerated development."

William Pecora, who alienated Santa Barbarans as director of the Geological Survey, was promoted to undersecretary at Interior in 1971, but that career was cut short when Pecora died in July 1972.

The Interior Department went through three secretaries in the first half of the 1970s. Rogers C. B. Morton was probably the most independent of the oil industry, but he was involved in Nixon's "accelerated development" response to the 1973 OPEC embargo. Under President Ford, secretaries Stanley Hathaway and Thomas Kleppe continued the Nixon initiatives. On balance, these changes appeared to favor the oil industry. But maybe changes were ahead.

In California, Jerry Brown became governor in 1975. His predecessor, Ronald Reagan, had played a muted role in offshore oil activities, leaving those problems to his attorney general's office, the State Lands Commission, and other state agencies. Brown came in with a strong antipathy toward offshore oil activity.

Before the end of 1976, Jimmy Carter had defeated Ford, and had named Governor Cecil Andrus of Idaho as Interior secretary. Andrus would arrive with an environmental reputation, and might—with Carter's blessing—steer a new course.

But the retiring, unelected Ford administration had one final gasp. The interim between a lost election and the last day in office is an administration's last window of opportunity. As we have seen, in 1974 the retiring Reagan administration in Sacramento approved permits for drilling forty-eight new wells from platforms in the channel.

Two days before the Ford administration was to leave office, the Interior Department issued a map showing 1,141,818 acres of waters off California for potential leasing in sale 48, still more than a year away. Of the 217 tracts involved, 105 were in the Santa Barbara Channel, covering almost everything not already under lease there. Many lease tracts intruded into the shipping lanes. William

Grant, head of the Bureau of Land Management's Pacific office, asserted these sensitive areas were selected for their production potential. Secretary Kleppe indicated the channel's unleased areas contained forty million to fifty million barrels of recoverable oil, in addition to twenty billion to fifty billion cubic feet of natural gas. Most of the government's information came from the oil industry. Again, the "market," as defined by the industry, was dictating energy policy.

Other proposed lease tracts were in the San Pedro Channel off Los Angeles, offshore Huntington Beach and San Diego County, and about fifty tracts south of San Nicolas Island. Grant said that some leases could be eliminated from the sale, but Bill Press, director of Governor Brown's Office of Planning and Research, described the lease map as:

> ...the outgoing administration's attempt to lock something into motion before the new team is aboard....Interior has ignored or overruled every concern or objection expressed by California representatives over tract selections, lease sales, exploratory permits or development permits.

Mayor Pete Wilson of San Diego protested leasing areas near that city's waterfront. He was supported by the San Diego Comprehensive Planning Organization, representing the county of San Diego and its thirteen cities. In a letter to Andrus, the group protested "the lack of a coherent energy policy that would demonstrate a need for the oil in light of the probable serious and adverse impacts on local economies and environment."

The Carter-Andrus Years

Three weeks after taking office, Andrus assured the House Interior Committee that he would thoroughly review the Ford administration's leasing plans and would feel free to make changes.

In May he postponed Southern California lease sale 48 from March 1978 until 1979 or later, and the Central and Northern California sale 53 from October 1978 to an indefinite date. It might even be dropped. He said production was needed more in the Gulf of Mexico where it could serve the central states, than in California where Alaska North Slope oil would soon be arriving.

Californians began to feel that for the first time they had the ear of the Interior Department on offshore oil leasing. The feeling was reinforced in June when Andrus met in Ventura with officials, activists and the press from the two counties facing on the channel.

Andrus postponed the Southern California lease sale after reviewing the Ford administration's proposals. "When I looked at the lease tract maps and the shipping channels, I just could not call the sale until we had more answers," he said. "I could not in good conscience ignore the input of the state and communities in asking for a delay in lease sale 48." He emphasized repeatedly that he wanted energy impact problems worked out on the local and state levels in order to avoid federal decision-making on local impact issues as much as possible.

> I wish that the Department of Interior had not stuck its nose into your business some time ago....I wish that you had worked it out among the corporate structures and the state and the citizens of the county and we wouldn't be in the rhubarb we're in....For whatever reason, the Department of Interior made a mistake.

It was assumed he referred to the conflicts between local and federal officials over the lease sale of 1968, the blowout and other ensuing events.

> Events of the past aside I can assure you that in the future the Department of Interior is going to listen to the supervisors of the counties. We're going to listen to the representatives of the state of California. We're going to listen when the governor [Jerry Brown] of your state writes us a letter and requests and then demands that we do certain things. I don't blame him. I came from a gubernatorial chair [in Idaho] and I know how absolutely frustrating it can be when the feds run right over the top of you. It just drives you up the wall.

In June 1977 Andrus invited all interested parties to submit comments on a revised three-year leasing schedule beginning in 1979. Two months later he moved lease sale 48 from March 1978 to June 1979, and lease sale 53 from October 1978 to February 1981.

That was encouraging, but it was not just federal leasing policies that disturbed coastal communities. They felt that federal regulation of drilling, production and transportation in the Outer Continental Shelf was equally irresponsible. For instance, Exxon not only obtained a federal permit for its big Hondo production platform, but was allowed to anchor an offshore storage and treatment plant nearby where tankers could take on oil without interference by state and local authorities who viewed tankers as air-polluters and maritime hazards. The oil industry and the Interior Department rejected state and local assertions that air pollution from platforms in federal waters seriously degraded air quality onshore. That dispute would not be resolved for seventeen years.

Andrus agreed that oil processing onshore under state and local controls was preferable to processing, storing, and tanker-loading several miles offshore. But in August his department concluded that it lacked authority to withdraw the permit for the offshore storage and treatment plant.

Not only would Exxon carry its oil away in marine tankers, but as North Slope oil entered the eight-hundred-mile Alaska pipeline, the Coast Guard announced that tankers carrying that oil from Valdez would pass through the Santa Barbara Channel instead of outside en route to the Los Angeles area or the Panama Canal. Shippers preferred the sheltered waters of the channel, and the Coast Guard believed that moving through the channel's established (but not mandatory) shipping lanes would be safer than navigating outside the channel where there were no vessel lanes.

Santa Barbarans, reading that thirty tankers had spilled 1.3 million barrels of oil worldwide during the first nine months of 1977, wanted as little as possible to do with the oil carriers. But Coast Guard officials insisted that vessel traffic through the Santa Barbara Channel was too slight to pose any danger. In offshore oil matters, California had never relied heavily on protection from the Coast Guard which became increasingly allied with the oil industry. At some hearings, Coast Guard testimony exceeded concern for safe and clean operations, and became partisan lobbying. The Coast Guard was not required to be seriously concerned with resource protection, but neither was participation with anti-environmental partisans part of the job description.

Help was sought elsewhere, and it was promised by the Interior Department. Before the end of Carter's first year, however, the California honeymoon with Andrus was beginning to end. The Brown administration in Sacramento complained that assurances of cooperation with state and local governments had not materialized. In a draft report in February 1977, the governor's Office of Planning and Research charged that the federal government planned offshore oil lease sales without sufficient information and failure to allow the state to participate in the process until most of the decisions had been made. The report was based largely on the record from the Nixon-Ford years, and before Andrus had assured Californians that things would be different. But the final version issued in December was substantially the same as the earlier draft. Not much, if anything, had changed, the governor's staff reported.

Local officials and citizens groups agreed. Interior's Bureau of Land Management conducted hearings for two days late in 1978 on an impact report for a Southern California lease sale. Most of the testimony was not on the impact report, but on whether the sale should proceed. As far as Santa Barbara County was concerned, it did not make much difference whether the testimony was relevant; federal agencies were not listening anyway.

"We cannot responsibly expend time and money on a federal procedure in which we are encouraged to give our views so they can be taken into account, only to have those views summarily dismissed," the Board of Supervisors wrote over the signature of Chairman Robert Kallman. The letter was read by Al Reynolds, the county's environmental resources director, who added on his own:

> The report never points out that a national conservation program could more than compensate for any oil foregone temporarily from withdrawing this sale.

Even the federal Marine Mammal Commission urged the Interior Department to postpone the sale pending more research on how it would affect marine mammals, especially the increasingly rare Guadalupe fur seal. The Bureau of Land Management replied that it would be careful, and if any mammals appeared to be threatened, leases could be cancelled or sales could be postponed. Anyway, the BLM maintained that the Fish and Wildlife Service and the National Marine Fisheries Service had not recommended postponement. While Andrus declined to cancel any sale, he did withdraw parcels from time to time, usually for environmental or safety reasons.

Meanwhile, Congress finally got serious about reviewing the leasing procedures in the 1953 Outer Continental Shelf Lands Act. During the 1969 Santa Barbara Channel oil spill, Secretary Hickel repeatedly described the act as grossly out of date. But not until 1978—nine years later—did Congress amend the act. The amendments required greater Interior Department cooperation with state and local governments in offshore oil leasing, and directed the department to publish five-year leasing programs, starting with 1980. Andrus went to work on this, and in March 1979 proposed a schedule of twenty-six lease sales for 1980-1985, including four off California.

That view of the future fell hard on California which was still fully engaged in protesting the previously-scheduled June 1979 Southern California sale. At a California Coastal Commission hearing in March, a dozen public agencies urged that the sale be postponed, curtailed or conducted under more rigid environmental safeguards. Except for the oil industry, every group who testified wanted the sale curtailed or scrapped. All the usual arguments were raised—resource degradation, maritime hazards, lack of need for low-grade oil, and the risk of inducing more state leasing. But the issue that stuck was air pollution.

Speaking for Santa Barbara County, Al Reynolds told the commission that the Interior Department had not yet published its air pollution regulations for the sale. Coastal communities needed assurance that offshore operations would not frustrate their efforts to meet federal Clean Air Act standards. The commission agreed, and asked Interior to postpone the sale until it demonstrated how air pollution would be held to acceptable levels.

Preliminary federal regulations were published, and the Interior Department acknowledged that the state had authority to veto plans for OCS operations that seriously threatened onshore air quality. That would seem to have settled the air quality conflict—over oil industry protest—but the dispute persisted for years because the department, the industry, the state, and local agencies could not agree on how much air pollution from the offshore platforms came ashore nor on a method to find out.

Despite legal challenges, lease sale 48 occured in June as scheduled. It was the first offering in the Santa Barbara Channel since the 1968 sale that led to the blowout. Although 148 tracts were offered between Point Conception and the Mexican border, the Santa Barbara Channel again was the major attraction. Fifty-five tracts received bids, thirty-two of which were in the channel. The sale brought high bids of $527.8 million, $110 million more than the 1975 Southern California sale that did not include the channel. The highest single bid was $95,475,000 for a tract off Point Conception, a hint of what was to come in a region of storms, maritime disasters, and no history of oil production; the Bermuda Triangle of the West Coast.

Andrus Tarnishes

Now well into his third year of the Carter administration, Interior Secretary Andrus was losing luster rapidly with the state of California and coastal environmentalists. He did not plan it that way, but he was being digested by a bureaucratic system that kept things pretty much as they had been. Even if the secretary had concern for local interests, the voices from that level were strained through bureaucratic filters that diluted the real ardor before it could reach Washington.

Beyond that, top management at Interior had its own techniques to cut the states and coastal communities short just when they thought they had some statutory protection against the heavy hand. Nowhere was this more apparent than at meetings of the Pacific Regional Technical Working Group, a sub-unit of the national Outer Continental Shelf Advisory Board, established by the secretary in 1975. The regional group included members from state and federal agencies, local governments, the oil and fishing industries, and environmental organizations. During this period the chairs were William Grant, head of the Bureau of Land Management's Pacific Outer Continental Shelf office and William Northrop, executive officer of the State Lands Commission. Each represented the agency responsible for offshore leasing in his jurisdiction. They agreed on almost nothing.

An intensive conflict centered on the drafting and crafting of Secretarial Issues Documents. These were reports prepared by the regional BLM office—and reviewed in Washington—to guide the Interior secretary in deciding on the size, location and timing of offshore oil lease sales. Before a SID was prepared there were extensive public hearings on the lease sale and its potential impacts. But little of this input was incorporated in the document. Conversely, much was included that was not discussed at the hearings, such as federal revenue benefits, social values, economic benefits to the communities, and other data generally favorable to the sale and injected by the regional BLM office.

This report was not subject to public review before the secretary decided on the lease sale details. It was strictly an in-house report. It was available, on request, *after* the decisions had been made. The first major confrontation over this issue erupted on August 7, 1980, when the Pacific Technical Working Group spent five hours dressing down the Pacific BLM office for the way lease sale decisions were made. The dispute continued at several subsequent meetings. The group urged that the reports be reviewed by the Technical Working Group before they go to the secretary, or at least before he makes his leasing decisions. "The SID [Secretarial Issues Document] should incorporate in the initial discussion a summary of public testimony and comments on lease sale 53," the group reported in a 1980 resolution.

Sale 53 was set for May 1981. During two weeks of hearings—from 8:00 A.M. until late at night, from Santa Rosa to San Luis Obispo—testimony was almost unanimously against the sale. The Technical Working Group wanted the secretary to get the flavor of public hearings. It wanted twenty-one issues added for discussion in the report to Andrus. They were concerns expressed at hearings but not conveyed to the secretary in the Secretarial Issues Document. The group protested that the bases for conclusions in the report to the secretary were often missing. Too much rested on unexplained assumptions.

Protective Wall

In private muttering within the California state bureaucracy, the name of one federal official was constantly cited as a wall between the public and the Secretary of the Interior. With remarkable survivability she was serving her third administration and fourth interior secretary. Heather Ross, Deputy Assistant Secretary of the Department of the Interior for Policy, Budget and Administration, was reputed to be the chief censor and editor of offshore leasing information before it reached the secretary.

Many environmentalists and staff aides to the California Coastal Commission, State Lands Commission and the governor's Office of Planning and Research believed she was the nation's most influential person on offshore oil leasing policy, and that she leaned heavily toward industry interests. Heather Ross' name surfaced in public discussion August 8, 1980. Warner Chabot, a planning consultant to Mendocino County, asked Grant:

> Is the indomitable Heather Ross to do the last editorial interpretation of the SID? I fear if she does, it may not be an accurate representation.

"She will be among those, yes," Grant replied. "That's her job as deputy undersecretary."

Our paths finally crossed one day in a Los Angeles federal courtroom, and I requested a few minutes of Ross' time. She was reluctant, but I stood between her and the exit and inquired about her reputation. Professing innocence, she implied that she was not much more than an Interior Department clerk who transmits California's reports to the secretary "just as they come to me." She did not edit the information, she said; she coordinated it. "The state's position gets full consideration," she said, and eluded further questioning.

The Technical Working Group also feared that Grant was not adequately conveying its concerns to Washington. Santa Barbara County officials protested that some of their recommendations to the Bureau of Land Management got lost, and Humboldt County officials said some of their recommendations were misinterpreted in BLM reports. But Grant repeatedly warned the Technical Working Group that it was exceeding its purview in many discussions and resolutions. John Fields, Grant's assistant, "noted that recommendations of the RTWG [Regional Technical Working Group] should be oriented towards scientific and technical input, rather than towards inputs pertaining to policy areas."[3]

This failed to deter the group. At one meeting, for instance, it recommended that reducing the need for petroleum through conservation, efficiency and the use of alternative fuels be as much a part of the nation's energy policy as increasing oil production. To emphasize its lack of confidence in the customary channel to the secretary, the group indicated that Northrop, the group's California co-chair, would deliver the recommendation rather than entrust it to the regional BLM staff.

Grant often regarded the group as obstructionists, and said so. He told the group he was "turned off" by excessive demands of state and local governments and environmental groups. The "credibility" of the group was damaged by the "irresponsible" views it expressed, Grant said.

These "irresponsible" views were held "almost unanimously by every board of supervisors along the northern California coast," replied Richard Charter of the Oceanic Society. "They're damned mad," he said, over plans to lease in sensitive wildlife habitats, among fragile coastal resources and in commercial fishing grounds. But Grant insisted that "the state cannot justify the position it has taken" in urging deletion of four northern offshore basins from lease sale 53. Grant's attitude made it difficult for Californians to accept him as the bearer of the state's case to Washington. (Andrus subsequently deleted all but the Santa Maria Basin from the lease sale.)

At the end of the two-day meeting, the Technical Working Group adopted resolutions that charged the BLM with developing a lease area ranking system "for industry interest and resource potential, but not for environmental information and multiple use conflicts," and recommended against reintroducing areas previously omitted from sales in the same region.

By this time the Reagan administration had taken office, and the leasing process was largely unchanged from the Carter and pre-Carter years. Like a cargo-laden tanker, the bureaucracy proceeded with a ponderous momentum and could be nudged into a new heading only long after the order to change course.

Congressional Nudge

Complaints by the Technical Working Group were repeated in August 1979 hearings in San Francisco before the House Select Committee on the Outer Continental Shelf. For two days California and local officials and citizens groups protested that the Interior Department ignored or slighted their concerns and had not justified the ambitious quest for low-grade oil off California.

The committee, chaired by Representative John Murphy of New York, was convinced of the need for change. In a letter signed by eight committee members, Murphy urged Andrus to get in closer touch with state and local governments, especially on lease sale 53 scheduled for 1981. Murphy said the committee was impressed with the unanimity of state, local and environmental witnesses in their opposition to the sale "and their belief that they had not been afforded the chance to make meaningful submissions regarding the leasing decisions involved in the sale."[4]

Andrus replied that he had been unfairly accused by state and local officials, asserting he had conferred closely with them in arranging lease sales. But in January 1980 when the Interior Department announced the area considered for lease sale 68, state and local governments did not see it that way. The sale area between Point Conception and the Mexican border encompassed sixteen million acres, including the ecological preserve seaward of the state sanctuary off Santa Barbara, and the federal buffer zone seaward of that. Also up for reconsideration were Santa Monica Bay and other areas that had been withdrawn from earlier sales. Seeking to calm the outrage, officials at the Bureau of Land Management said that "historically, outer continental shelf lease sales. . .have actually offered only a minor fraction of the vast areas included in the initial call."

But coastal communities wanted to know why local governments were forced to repeatedly defend the same areas with each new lease sale proposal. Community wishes were known, but were put to the test time after time by a federal administration that had promised close cooperation with coastal governments.

Santa Barbara County supervisors wanted to confront Andrus about this, but instead they met in Los Angeles with Undersecretary James Joseph. The press was excluded, but officials who were there said he told them that "talking to me is the same as talking to the secretary."

"Santa Barbara appears to be treated like the Indians," City Councilman Hal Conklin said after the meeting. Joseph had conceded nothing. "They're undermining the very system they boast of," said Robert Hedlund, chairman of the Board of Supervisors.

Joseph told the demoralized delegation that the Interior Department believed there is oil in the fifteen tracts comprising the ecological preserve and the buffer zone which were established by executive order in 1967 and 1969. Talking with reporters after the meeting, Joseph said that "the executive order needs a new review....The area needs to be looked at." It was not clear whether he referred to both the ecological preserve and the adjacent buffer zone or just the latter. But lease sale maps indicated both were to be considered for leasing. In July the Interior officials said they would respect the preserve next to the state sanctuary in the Santa Barbara

Channel, but not the buffer zone. It also backed off of Santa Monica Bay, and withdrew 72 tracts that the Pacific Missile Range command said would interfere with its military operations. But more than 220 tracts remained, and would be bitterly disputed before the sale in June 1982.

Meantime, however, was the matter of lease sale 53 set for May 1981 off Central and Northern California. Northern California fishers had been especially vocal in protesting oil operations. In October 1980 Andrus announced that the waters off Northern California would not be part of the sale. All that was left of the sale, then, were the waters off Santa Barbara and San Luis Obispo counties. The tract selection was scheduled to be announced late in November, but many observers saw a political motive in announcing the reduction in the sale area before the presidential election. It did not please the two remaining counties, but it undoubtedly gained votes along the Northern California coast.

"Reoffering" Sales

In the meantime, the Interior Department expanded its five-year leasing plan to include thirty-six offshore sales instead of twenty-six as earlier proposed. The new plan not only rejected requests for delays in the four California sales, but added "reoffering" sales which would include tracts offered but not sold at previous sales. The result would "double our trouble," said Deni Greene of Governor Brown's Office of Planning and Research.

After announcing he would quit in January 1981 regardless of the election results, Andrus braved a Santa Barbara audience less than two weeks before the election. Noting that every Pacific Coast lease sale from 1968 through 1984 would impact Santa Barbara County, a reporter asked whether the county was not being asked to contribute more than its fair share to the nation's energy supply.

"No," he replied. "We cannot control where oil is found. Oil must be produced where it is found, if it will not cause undue impacts on the adjacent communities...." He said oil activities need not degrade the county. Coastal Californians were not so sure.

The drumbeat of complaints about oil leasing was of course not the only problem besetting the Interior Department or the Carter administration as the 1980 election approached. Topping the inventory of dilemmas were the fifty-two hostages held in Iran. But the public perception of Interior's offshore oil leasing program was one of the irritations that set Andrus on edge.

In a brief news conference in Santa Barbara in October 1980, I asked Andrus if he was aware of complaints that he was shielded or isolated from public opinion on offshore leasing through a series of protective devices in the system. Clearly irritated by the question, Andrus replied that such criticism is unfounded. "I read the hearing records. I read the papers from the coastal areas. I am kept advised by field representatives in the Interior Department."

That was about it, and it did not answer any questions. "Field representatives in the Interior Department" were, according to the California critics, part of the problem, not the solution. What the "hearing records" contained was not clarified, nor was how much time he had to read "papers from the coastal areas."

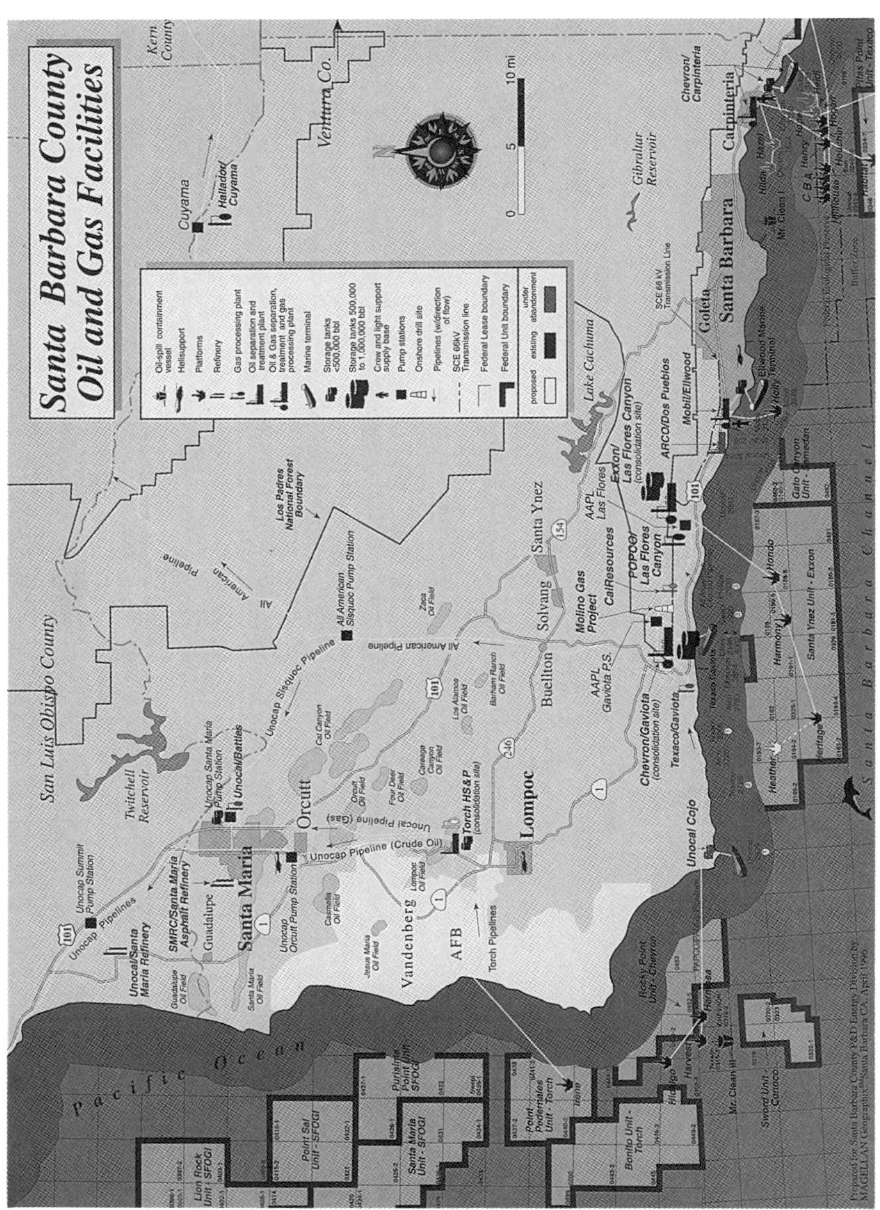

Santa Barbara County Oil and Gas Facilities. (Santa Barbara County Division of Energy)

Chapter Seven

"Frozen Fire" Plan Defrosts

While Californians were trying to cope with aggressive offshore oil leasing, another behemoth was reconnoitering the state's coast—liquefied natural gas, or LNG. Few know what LNG was, or could become if it encountered a spark. The first news stories (including mine) referred to *liquid* natural gas, which is quite different from *liquefied* natural gas. So public response was only tepid in 1973 when Southern California Gas (a subsidiary of Pacific Lighting) received a permit to conduct geologic testing for an LNG tanker pier at Port Hueneme, and then applied for permits for three Southern California LNG terminals. The main public concern then was increased tanker traffic in the Santa Barbara Channel.

The gas utility created the Western LNG Terminal Company (later joined by Pacific Gas & Electric) to operate ports at Little Cojo Bay near Point Conception, the western wild-weather entrance to the Santa Barbara Channel; at Oxnard, near the Port Hueneme deepwater port at the eastern end; and in the Los Angeles harbor. Each would have berths for two LNG tankers, four 550,000-barrel LNG storage tanks, and equipment for reconverting liquefied gas to its natural state for distribution to consumers. Tankers would carry the liquefied gas from Indonesia and Alaska.

The alarm was sounded in January 1975 by a *Los Angeles Times* reporter who wrote that "LNG has a potential for danger that is almost unthinkable when spilled on water."[1] He quoted Dr. David L. Jaquette of the Rand Corporation as saying that LNG spilled on water vaporizes very quickly and can create a foglike cloud subject to being blown by the least breeze.

> The cloud would be deadly in itself....It would asphyxiate and freeze anyone in its path. But the most important danger is igniting the cloud. Any ignition would cause a flash back all the way to the source.

LNG is not the natural state of natural gas, and should not be confused with the more benign natural gas liquids (NGLs) such as propane and butane, or liquefied petroleum gases (LPGs) that are marketed as bottled gas. They too can explode, but the eruptions are puffs compared to the energy of escaping LNG if it ignites. LNG is natural gas "chilled" to 260 degrees below zero Fahrenheit (minus 162 degrees Celsius). That reduces the volume to .06 of its gaseous state. It is called a "Time Bomb" in a book of that name published in 1978 by Peter van der Linde and Naomi A. Hintze, and "Frozen Fire" in the title of a 1979 book by Lee Niedringhaus Davis. Tankers carrying such cargo are sophisticated vacuum bottles or floating doomsday bombs, depending on one's point of view.

As 1975 wore on, coastal Southern Californians saw ever more clearly the ramifications of LNG port plans. In September, the California Resources Agency warned that the Los Angeles harbor was no place for LNG tankers. Even though a tanker accident was unlikely, "in a populated area such as Los Angeles harbor, this risk is substantial," said Claire Dedrick, Sierra Club activist whom Governor Jerry Brown had just appointed state resources secretary. Unimpressed by the warning, a Los Angeles City Council committee voted on December 17, 1976, to let Pacific Lighting Corporation begin work on an LNG terminal in the harbor. The next day the oil tanker *Sansinena* blew up in the harbor when gas vapors ignited over a cargo tank. The blast broke windows 20 miles away, and was felt twice that distance from the harbor. Sobered and shaken, the city council denied the LNG permit, but six months later the potential revenue loomed larger than the receding memory of the explosion, and the LNG terminal project was approved.

But apprehension mounted elsewhere. A study by Science Applications, Inc., warned that a tanker leak of 125,000 cubic meters of LNG at Oxnard could jeopardize 50,000 people. Some reviewers faulted the report, asserting that before the cloud could envelope that many people, it would reach a spark or flame and would quickly burn back to its source. The resulting fire storm or explosion, however, would hardly have been a benign alternative.

One could chose, however, from widely varying conclusions of the many consultants on the case. Santa Barbara County officials were not consoled by federal or industry experts, and became increasingly uncomfortable about the prospect of an LNG port nearby. They said nobody knew the full potential of a major LNG accident, but they believed that a Federal Power Commission (FPC) report seriously underestimated the risks.

"The potential danger from an LNG fire is probably the largest ever presented to modern society, especially in the case of the three combined facilities" at Point Conception, Oxnard and Los Angeles, said Al McCurdy, a Santa Barbara County environmental specialist. Albert Reynolds, head of the county's Office of Environmental Quality, repeatedly urged the FPC to schedule California hearings on the three proposals. The California Energy Commission and Public Utilities Commission extended similar invitations. But the federal agency declined to leave Washington.

That did not mean there were no hearings in California. At least a dozen city, county, state and federal agencies deliberated for two and a half years over places for an LNG terminal. It came down to five sites: two near nuclear power plants, one on an earthquake fault, two in urban areas, one on a military base, and one on sacred Chumash land. The total is more than five because some sites had more than one of the disadvantages.

As the California Coastal Commission and the California Public Utilities Commission shuttled up and down the coast, they heard almost nothing but protest against any site. The sessions were mostly unproductive. For one thing, as William Ahern of the Coastal Commission staff said:

> We in no way are saying that these five sites are great or wonderful or favorable....California is a terrible place to find sites, where earth [tectonic] plates are moving one against the other....So far all we [staff] have done is to say it is possible physically to build these damn things on these sites.

In other words, there were no safe places, and the site-rating system was in chaos. Some agencies shifted their ratings as new information came to light, and some agencies went against the recommendation of their staffs. Hearings were frustrated also by the dispute over who had the final word in selecting a location. The United States Department of Energy was created on October 1, 1977 and assumed control on the federal level. But that did not settle the jurisdictional dispute among the county, state, and federal agencies.

The turf strife among state agencies was settled in an agreement between Governor Brown, the utilities, and key legislators. State Senator Alfred Alquist introduced SB 1081 which would virtually assure the selection of Point Conception. Under the bill, the deadline for making the choice eliminated every other option because impact reports could not be prepared in time. Only the Point Conception report would be completed.

The bill specifically eliminated Oxnard and Los Angeles because they were densely populated areas. It took siting authority away from Santa Barbara County and the California Coastal Commission. The legislation required the State Public Utilities Commission to make the decision.

(Attitudes among key players were taking an unusual twist. Governor Brown, who was leading the fight against offshore oil activity and for coastal resource protection, now assumed a pro-development stance to shed an anti-business aura. Even Sierra Club leaders, convinced that an LNG port was inevitable, joined site selection discussions rather than oppose the entire LNG notion. Meanwhile, staunch friends of industry such as Edward Teller, the Rand Corporation, and Dorill Wright were wary of—or opposed to—an LNG terminal. Wright, mayor of Port Hueneme and a member of the Regional Coastal Commission, did not want the fearful business next door in Oxnard.)

By the time the Alquist bill came before the Assembly Resources Committee in August 1977 the agreement was virtually airtight. Alquist threatened to remove his name from the bill if weakening amendments were attached, explaining that the gas companies would not accept changes. The utilities seemed to sit as the third house of the legislature. The bill glided through the committee on a nine to two vote, passed both houses of the legislature and was signed by Governor Brown.

About this time, environmental groups began asking whether the governor had a personal financial interest in LNG importation. The link would be through his father, former Governor Edmund G. (Pat) Brown Sr. The elder Brown had been an officer and thirty-one percent owner of U. S. International Investment Corporation, a holding company that owned Perta Oil Marketing Corporation, the sales agency for all Pertamina oil sold in the United States. Pertamina was the Indonesian government oil company. Pat Brown maintained that he gave his thirty-one percent interest to his children—all but Jerry. Anyway, he noted, Perta Oil dealt only in crude oil and had nothing to do with liquefied natural gas. The former governor said he was eager to see the LNG project proceed, but that no one in the family had a financial interest in it.

He acknowledged that Perta had enhanced Jerry Brown's 1974 gubernatorial campaign by $21,000, but that when he (Pat Brown) learned of it, he told Perta to stop contributing to Jerry's campaign.

That issue blew over quickly, and the bizarre site evaluation continued. While in reality Point Conception had been chosen, the State Public Utilities Commission and California Coastal Commission scheduled make-believe site-selection hearings. Everywhere they encountered overwhelming opposition to LNG industrialization. The limitations imposed on state and local agencies reduced the hearings to charades. "The whole energy development planning process has gotten out of control," said William Ahern, energy coordinator for the Coastal Commission. Even within the Public Utilities Commission—the only agency the legislature empowered to select the site—progress was impeded. "[The hearings] have become terribly complex and often are reduced to attorneys cross-examining each other while you sit on your rear end waiting to get in a word, perhaps, for solar energy or energy conservation," Ahern said.

But he urged interested persons to make use of the forums that were available, including hearings of the Santa Barbara County Planning Commission whose only remaining LNG function was to recommend conditions for the Point Conception plant. The only person to "violate" the ban on discussing the merits of the project was Kote Lotah, representing the Chumash community. Point Conception has archeological remains, and the Chumash regard the promontory as the souls' point of departure into the next world. "We're going to fight LNG all the way," Lotah said. "We will not discuss conditions for bringing it here. We're going to use all the laws and resources available to fight it."

After twenty-five hours of hearings in March and April, County Planning Commissioner Joan Wells moved that no site be selected in 1978 because there was not time to consider all the ramifications. The commission chair and county lawyers interrupted to remind Wells that the motion was out of order; beyond the scope of the hearing. Legal or not, four other commissioners supported the motion which failed on a five to five vote.

Wells' "out-of-order" motion was not considered outrageous by the Board of Supervisors which supported a bill by Assemblyman Hart that would delay a PUC permit decision for one year.

This would undermine the Alquist bill that expedited a permit for the Point Conception site. Hart's bill stood no chance of passage.

Wells continued her protest in the final Planning Commission vote on land use conditions for the project. Now commission chair, she said in August 1979, "I will abstain because I think it is outrageous that an LNG plant is getting closer to [being approved for] Point Conception. This is a protesting abstention."

This was not the only rear-guard action against the selection of Point Conception. The California Coastal Commission recommended Camp Pendleton near San Diego, and the Federal Energy Regulatory Commission (FERC) staff preferred Oxnard. When it came down to the decision-makers, however, Point Conception was selected by the state PUC in July 1978, and by FERC in September 1979. That avoided a showdown over whether the state or the federal government had permitting authority.

But protests continued, and other events cast a pall over the project. On October 6, 1979, an LNG regasification plant exploded at Cove Point, Maryland, killing one man and demolishing the cinderblock building. Along with this, the price of natural gas dropped and the anticipated shortages never materialized. Pacific Gas & Electric, one of the two Western LNG Associates partners, withdrew its financing for importing gas from Alaska.

The project was "virtually dead," said Robert Hight of the State Lands Commission legal staff. A spokesman for the LNG group insisted that the project was still very much alive. But before the end of 1982 John Sproul, Pacific Gas & Electric executive vice president, announced that construction was more than two years off because there was no need for more natural gas sources before 1985.

Before 1985 rolled around, however, the PUC ruled that there was still no need for LNG and that the plant permit was "no longer applicable." It also ruled that utility customers would have to pay $144 million over four years to help pay for the cost of carrying the project this far.

In 1986, fourteen years after filing the LNG permit application, the two utilities withdrew their permit application, which the PUC had in fact done for them in 1984. In 1987, after ten years, the California legislature repealed the skillfully-crafted and craftily-enacted SB 1081.

Chapter Eight

Reagan and Watt Push Forward

Although he continued the "accelerated leasing" program started by Richard Nixon, President Carter offered Southern California some protection from the encroachment of the offshore oil industry. Under the Marine Sanctuaries Act of 1972, Carter set aside 1,252 square miles of ocean around the Santa Barbara Channel Islands as a National Marine Sanctuary.[1] The state had already prohibited leasing three miles seaward around the islands. The national sanctuary extended six miles seaward. The 1972 federal law granted the president authority to create marine sanctuaries without congressional action. All he had to do, technically, was sign a paper. But in this case the process was tumultuous, and it almost failed.

Santa Barbara County, the California Coastal Commission and environmentalists wanted a sanctuary to encompass the entire channel. The National Oceanic and Atmospheric Administration (NOAA), which manages the sanctuaries program, scheduled hearings on the request in Santa Barbara. The NOAA hearing team was young and only two small sanctuaries had been established under the 1972 act. Inexperience contributed to the ensuing chaos.

The NOAA panel was caught between formidable and hostile coalitions. The oil industry and the Coast Guard, for different reasons, protested any plan for a sanctuary. Even more strident were commercial fishers. A group that is almost impossible to organize for any purpose, the fishers this time charged in, united and shrill, condemning any hint of more restrictions on their industry. Somewhat intimidated, NOAA agreed not to add new fishing rules.

But the fishers did not trust the bureaucrats, and even after the sanctuary was in place and they had won their point, they continued their protest. They achieved by shouting what the oil industry could not accomplish through litigation. Twice the Western Oil and Gas Association, the major petroleum trade group on the Pacific coast, went to court trying to prevent the loss of potential leasing areas, and twice it lost.

Not that the county and its supporters received everything they wanted. NOAA officials indicated they could not bite off a chunk as large as the entire channel. After two years of wild hearings, NOAA settled on a sanctuary extending six nautical miles[2] around the five islands, still an imposing expanse of water. New oil leasing was banned, but the rules allowed exploring and producing in already-existing leases within the sanctuary.Carter said as he signed the designation on September 21, 1980:

> The area clearly deserves marine sanctuary status....The islands and surrounding waters are an exceptionally productive ecosystem. They provide feeding and breeding grounds for one of the largest and most varied assemblages of seals and sea lions in the world. They are one of the richest resource areas in the United States for marine birds, including the endangered brown pelican. The area has become particularly important as the pressures of human development have driven these species from one refuge after another on the mainland. It will complement the Channel Islands National Park that I recently approved."[3]

It appeared that the sanctuary had been delivered. But no. Before the regulations went into effect, Ronald Reagan came into office and suspended the regulations while the new administration studied their economic impact on private industries. It was part of a general freeze of pending regulations designed to carry out Reagan's campaign pledge to deregulate. The sanctuary regulations eventually were reinstated, but for a year everything hung in the balance.

The Interior Department and the oil industry had lost what they felt were potentially lucrative regions. But the Interior Department used it to dubious strategic advantage by threatening to offer oil leases within the sanctuary—despite the federal prohibition—and then withdrawing these areas later to demonstrate its willingness to cooperate with regional concerns.

The Five-Years Plans

The loss of the sanctuary for oil exploitation still left much of the Pacific Ocean to subdivide for leasing. The Interior Department and the oil industry made the most of it. Cecil Andrus' staff in 1980 drafted the first five-year plan which was challenged by California,

Alaska, the Natural Resources Defense Council and the North Slope Borough Community of Unupiat Eskimos.

The Court of Appeals in the District of Columbia agreed with much of the plaintiffs argument, and returned the five-year plan to the Interior Department for greater elaboration on environmental and economic impacts. In the meantime, however, the department could go ahead with the lease sales it had already scheduled. By the time the court ruled in October 1981, Andrus was back in Idaho and James Watt was nine months into his job as Interior secretary.

Watt had been director of the Mountain States Legal Foundation, a conservative Colorado law firm dedicated to private exploitation of natural resources without government interference. To a point—even *beyond* a point—he fit into Reagan's deregulation plan. Reagan pressed deregulation enthusiastically but without apparent rancor. Watt pursued it with strident audacity and religious ferver. God would provide resources indefinitely, he felt, for a rapacious society.

Watt visited the Santa Barbara Channel area in February 1982, touching down by helicopter on Anacapa Island in the Channel Islands National Park, and at the park headquarters in Ventura. Wearing a new plaid shirt with a button-down collar open at the throat, he described himself to reporters as "a very great secretary of interior" and a major asset to the administration. "I'm the biggest fundraiser they have," he said, responding to a question about whether he considered himself a liability to the Republican Party and the White House.

He might have added that he was the greatest membership boost some environmental groups had ever enjoyed. His approach to resource management was so alarming that membership ballooned in the already large organizations such as Sierra Club, Wilderness Society, and Audubon Society.

"Criticism has helped me become a very great secretary of interior," Watt continued, "because it has drawn the attention of Congress to Jim Watt and what he is trying to do. Congress has been very supportive." But it was largely congressional opposition that drove him from office eight months later. For the moment, however, he could blame his predecessor for the adverse ruling on the 1982-1987 offshore oil leasing plan. But not very convincingly.

Eager to promote private development of public lands, Watt had already proposed a new five-year plan, to start in 1982, that even more flagrantly ignored the concerns of environmentalists and coastal states and communities. Three scheduled offshore California oil lease sales were reset for earlier dates, and two were added to the new five-year plan. The plan would not only shorten the time between sales, but increase the area under study to eighty-three million acres.

Coastal communities protested. That was anticipated. But it went beyond that. The National Oceanic and Atmospheric Administration warned that unless there was more study of the consequences of the leasing schedule, "important management decisions will be made without the benefit of scientifically sound information."[4] The Coast Guard objected to leasing parts of Santa Monica and San Pedro bays because of heavy marine traffic. "Even oilmen fear the consequences of a headlong approach that yields little but time-consuming court fights," the *Wall Street Journal* reported.[5]

But Watt was undeterred. Less than a month after he took office he said that the four areas off Northern and Central California that Andrus had spared from leasing would be back on the bidding block. These were the Point Arena, Bodega Bay, Santa Cruz and Eel River basins. Coastal residents there wanted no offshore oil development, and in 1980 they thought they had won. Now it was back to the trenches.

Governor Brown and the California Coastal Commission protested Watt's plans, but in April 1981 Watt told Californians that their protests were ill-founded, and that the state had virtually no authority to interfere with the federal leasing program.

According to an Interior Department background paper:

> The controversy over offshore leasing is the classic problem of public interest versus local interests....Among the parties to individual sale decisions, local and state concerns do not reflect this overriding national need or perspective...There is...some modest risk of conflict with other ocean uses such as fishing, and of environmental damage from, for example, an oil spill, the probability of which cannot be reduced to zero....California argues that the risks are too great for the amount of oil to be found. This is false.

The report maintained that while Congress gave the states some opportunity to participate in offshore leasing discussions, "it did not, however, assign any decision authority to the states."[6]

This was a strong statement, but even more than anyone knew then, it set the stage for two immense conflicts that changed the course of federal offshore oil leasing in California. It underestimated the depth of the protest in northern coastal communities, and it raised the issue of the state's authority to interfere in federal oil leasing.

Before lease sale 53 in May 1981, Watt backed down and withdrew the four Northern California basins—as Andrus had done—but offered the Santa Maria Basin off Santa Barbara and San Luis Obispo counties. This included, among other resources, the range of the southern sea otter, still struggling to recover from near-extinction.[7]

During the opening of bids on May 28 attention shifted from the sea otter range to the waters off Point Conception in Santa Barbara County. But not until the last minute. While William Grant of the Bureau of Land Management opened and read 297 bids, 600 spectators at the Anaheim convention center responded with occasional murmurs. After two and a half hours, only four bids remained to be opened. Suddenly the crowd erupted with "wow!. . .wow!" as Grant read bids totalling $635 million from Chevron and Phillips Petroleum for four tracts west of Point Conception at the southern end of the lease sale area. It was nearly one third the amount of accepted bids for the entire sale of sixty tracts.

Chevron and Phillips knew something—or thought they did—that the rest of the industry did not. It was now clear how they got so smart. In the 1979 sale, Chevron and Phillips acquired leases that abutted the area of the 1981 sale. They made significant discoveries which they did not announce, and in 1981 they simply bought the adjoining tracts and added them to their initial holdings. They were at the western end of the Santa Barbara Channel where California "turns a corner" from north-south to east-west.

Santa Barbara Sandwiched

Reagan's first term, which exceeded Watt's tenure, was open season on offshore California. The pace was a sale per year, plus a reoffering sale in 1982 in which twenty-seven tracts were offered but none sold.[8] The state and coastal communities informed Washington there was no way they could plan for such a rapid succession of enormous auctions, but Reagan and Watt persisted and prevailed. Lease sale 68 off Southern California took place in June 1982, Sale 73 off Central California in January 1983, and Sale 80 off Southern California in October 1984. Each sale was ultimately whittled down from the enormous initial proposal, but cities and counties had to begin planning on the basis of the initial plans.

Just knocking parts out of each sale was an exhausting, time-consuming, costly exercise for local governments. The state, of course, was forced to respond to every offshore California adventure envisioned by the Interior Department. Santa Barbara County too had to cope with all four annual sales in Reagan's first term. Rich offshore oil deposits border the county off its southern and western coasts. The two shorelines meet at right angles at Point Conception, and that was the Interior Department's dividing line between Central and Southern California. Even when the Interior Department proposed a *Northern* California sale, the department combined it with *Central* California, and Santa Barbara County was again targeted.

State Demands a Voice

When pressed this severely, Santa Barbara County often turned to the state for help because in the late 1970s and early 1980s Sacramento assumed it had a voice in federal offshore oil leasing. The Interior Department disagreed, and the dispute ran for more than a decade up through the Supreme Court and back to Congress before it was settled in 1990. The federal Coastal Zone Management Act (CZMA) was part of the wave of environmental laws enacted in the early 1970s. It made federal funds available to states with approved coastal plans and gave states some say over what goes on in the Outer Continental Shelf where the states would otherwise have no authority.

The Coastal Zone Management Act required, among other things:

> Each Federal agency conducting or supporting activities *directly affecting* the coastal zone shall conduct or support those activities in a manner which is, to the maximum extent practicable, consistent with approved state management programs. (Emphasis added.)

In other words, the federal agency had to assure the state that the proposed activity would be consistent with the state's coastal plan. If the state and Interior Department did not agree, the matter could be mediated by the Commerce Department which administers the act.

The California Coastal Commission found lease sale 48 in 1979 inconsistent with state standards, but the Interior Department did not even argue that point. The department maintained that lease sales do not directly affect the California coast and consequently no federal consistency finding was necessary. A lease sale, after all, is only a contract, a piece of paper. It does not pollute or degrade anything.

The Coastal Commission disagreed, i.e., a lease sale sets off a chain of events which are difficult to modify after lease contracts are signed, and which can seriously damage coastal resources. Neither mediation nor an appeal to President Carter resolved the matter. The state was not yet inclined to sue, and lease sale 48 went ahead on June 29, 1979.

The state was aroused, however, by plans for lease sale 53 two years later because it included the sea otter range off Monterey and San Luis Obispo counties. Once considered extinct, the southern sea otter now was struggling to survive, mostly on abalone in the nearshore rocky-bottom waters. They were fervently defended as adorable creatures by coastal dwellers and visitors, and despised by commercial fishers who said the voracious otters were destroying the abalone fishery.

This time Governor Jerry Brown took the issue to court. Before Federal District Judge Mariana Pfaelzer the state argued that its coastal plan mandated protecting endangered species, and a major oil spill in the otter range could wipe out the entire population. Judge Pfaelzer ruled that Secretary Watt had violated the Coastal Zone Management Act by not first submitting the lease sale plan to the state government for its approval. In an order that took almost an hour to read, Judge Pfaelzer said that the act was not clear on the issue of "directly affecting" coastal resources, but that a thorough review of the legislative history of the act and its amendments clearly indicated Congress intended to require federal lease sales to be subjected to state approval. Viewing the ruling as a threat to the ambitious leasing plans, the Interior Department and the industry lost no time appealing.

More than that, the Commerce Department—under Interior Department pressure—rewrote the CZMA regulations to explicitly prohibit states from blocking lease sales. The new rule stated that lease sales do not directly affect coastal resources. It was not viewed that way by the House Merchant Marine Oceanography Subcommittee which gave notice it would draft a resolution of disapproval. In the Senate, the Commerce Committee indicated it too would review the matter. With congressional concern mounting, President Reagan withdrew the disputed regulation in November 1981.

(The same month a CZMA team visited California to determine if the state was adequately protecingt coastal resources. It was a periodic inspection required by the act. But many found it ironic that the administration that was undermining California's attempts to nurture its coast was now coming across the continent to verify that efforts met federal standards. In fact, the diligence of the California Coastal Commission was such that in 1987 Interior Secretary Donald Hodel said the commission was usurping federal authority. He asked the Commerce Department to investigate whether the California coastal plan should be "decertified" and denied federal funding, not because the commission was slighting its mandate but for taking it too seriously. Decertification would take away the state's authority to review *any* Outer Continental Shelf operations, not just leasing plans.)[9]

Three federal appeals judges spent more than a year pondering the meaning of two words: "directly affecting." They concluded unanimously, in August 1982, that federal offshore oil lease sales do directly affect state coastal resources, and that states had the right to review lease sales for consistency with state coastal plans. The frustrated administration appealed to the Supreme Court.

Nothing to date had done much to clarify leasing issues. On November 10, 1983, the California Coastal Commission listened to a 10-hour wrangle over lease sale 73 (plus a proposed *state* lease sale) that would include more than 2,500 square miles of ocean from Point Conception to Morro Bay. In the end, the commission rejected the offering, and went to court to stop the sale.

Michael Fischer, executive director of the commission, said the commission wanted offshore development extended over decades to slow the pace and reduce the impact, rather than operating it "like an Oklahoma land rush." The commission, joined by several environmental groups, charged the sale was a threat to sea otters, gray whales, marine life food sources, seals and sea lions, the Morro Bay estuary, and commercial fishing.

The federal district court banned the sale, the appellate court upheld the lower court, and Supreme Court Associate Justice William Renquist lifted the ban, all within three weeks. Maintaining the pace, the Interior Department opened bids for lease sale 73 on December 21—the day after Justice Renquist ruled—and almost nobody came. Only eight of the 137 tracts were bid on, and the bids totalled $16 million. "That's got to be the worst sale in history," said Art Spaulding, director of the Western Oil and Gas Association. The tracts offered were in deep water where production would be costly, and the price of oil did not make it worth the effort and expense.

If 1983 ended on a downbeat for the industry in the channel, 1984 dawned with a Supreme Court ruling that dismayed environmentalists, but elated the industry. In a five to four decision, the court ruled that the Coastal Zone Management Act gave states no voice in federal offshore leasing. Writing for the majority, Justice Sandra Day O'Connor argued that leases authorized companies "to engage only in preliminary exploration. Further administrative approval is required before full exploration or development may begin." Environmentalists and California officials said she did not understand the system.

If California's authority in this area was to be restored, Congress would have to amend the Coastal Zone Management Act to require that lease sales be consistent with state coastal plans. At least two California members of Congress were already drafting such an amendment. Congressional opposition had forced the Reagan administration earlier to back down on a proposed rule to eliminate state influence on OCS leasing. Maybe there was hope.

There was, but it required patience. It took Congress nearly seven years to restore state authority. When the CZMA was up for reauthorization in 1990, Congress simply dropped the requirement that lease sales must "directly" affect state coastal areas in order to be subject to state review and objection. The states were again legally reinforced for the struggle over offshore oil leasing. As it turned out, however, federal lease sales off California had largely run their course for the rest of the century.

Get Oil Out office in Santa Barbara was splattered with oil in 1975, presumably by opponents of the group's anti-oil activities. (*News-Press* photo by Steve Malone)

Chapter Nine

Reagan and Bush Back Down

President Reagan enjoyed the last year of his first term without the turbulence of James Watt. To help ease the way toward the 1984 elections, Reagan appointed his friend and fellow Californian, William Clark, to succeed Watt. Environmentalists hoped that Clark, who owned a ranch in San Luis Obispo County north of Santa Barbara, would empathize with coastal residents on offshore oil matters. During Clark's brief time at Interior, environmentalists got some of their wish. After Watt, it almost *had* to be better. "Clark...washed away much of the bad taste left by Watt," columnist Lou Cannon wrote in his book on the Reagan administration.[1]

Clark said he would overhaul the offshore leasing program, increase public participation, and give more attention to resource protection.[2] In revamping the Watt five-year plan, he placed all the scheduled California offshore lease sales on hold. In an address in Santa Barbara in May 1984, he maintained that the nation needed the oil off the California coast, but that it would be produced in consultation with concerned individuals, organizations, and above all with state officials. He was not available for questioning before or after his address at the University of California at Santa Barbara. He ducked in and ducked out.

During his tenure at Interior, Clark rescheduled the Southern California lease sale 80 for October 1984. He also eliminated two-thirds of the sale area proposed by Watt. It was still an enormous offering, covering 3,147,352 acres, centering again on the Santa Barbara Channel. As it turned out, the sale was nothing for Interior to cheer about. Of the 657 tracts offered, only twenty-five drew bids, most of them without competition. High bids amounted to $62.1 million.

It was not known then, but it was the last offshore California lease for at least the rest of the century. But at the time Interior officials were still planning more California offshore leasing, concentrating as usual on the Santa Barbara Channel.

"No other community in the United States is under so much oil pressure as Santa Barbara County," said Robert Knecht, former director of NOAA's Office of Coastal Zone Management. Knecht was familiar with Interior's OCS leasing practices. In February 1985, as a Santa Barbara resident, he was debating William Grant, head of the Pacific Coast office of the Minerals Management Service.

Knecht charged that Interior's leasing was coming too rapidly and with too little consultation with state and local governments. "The Interior Department has cut off grants to the states, has opposed oil royalty revenue-sharing, and in every other way has opposed assisting the states and local governments," he said. Recalling the dispute over the reach of the Coastal Zone Management Act, Knecht pointed out that the Interior Department had lobbied to reduce state and local voice in federal offshore leasing. "That doesn't seem like partnership to me," he said.

Grant insisted that the pace of leasing off Santa Barbara County had been "moderate, if not slow....We must proceed where the oil is....Even though we don't always agree [with Santa Barbara County], the situation is improving," he said, referring to the relationship between the Interior Department and the county.

A Hodel Five-Year-Plan

And so it seemed. Watt was gone and Clark had tempered the overheated leasing schedule during his brief tenure. But the appointment of Donald Hodel in 1985 to succeed Clark was not promising. While not as strident and bumptious as Watt, Hodel managed to stir up nearly as much contention. It should not have been surprising. Before coming to Interior, Hodel headed the United States Department of Energy where he resisted any serious efforts toward energy conservation and the use of renewable resources. Much later came news of the Energy Department's longterm mismanagement of the nation's nuclear energy program.

The new 1987-1992 federal oil leasing plan, announced in March 1985, included four offshore California sales. A scheduled 1986 offshore California sale, meanwhile, was moved to April 1987, and another was added later to the 1987-1992 plan. As with the two previous five-year plans, frequent changes made it difficult to know what the future held, or even what current policy was. As it stood, however, California could expect five more sales between

1987 and 1992. Santa Barbara County residents and officials urged Interior to delay more leasing until the county could accommodate production from previous lease sales.

Orange County supervisors, thoroughly familiar with coastal oil development, continued their opposition to federal leasing. The Board of Supervisors condemned the 1987-1992 five-year plan, and pressed for a National Oceanic Park between Catalina Island and Orange County "for the protection of our natural and economic resources." It wanted neither the petroleum pollution risks from offshore operations nor the air pollution, the drain on municipal services, traffic congestion and land use conflicts that stem from onshore support activities. A coalition of Orange County cities felt the same way.[3] The county's Environmental Management Agency reported:

> There is a growing and united interest among Orange County municipalities to challenge the Interior plan....All but one coastal city in this area have gone on record supporting establishment of a national Ocean Park in Orange County offshore waters....For local governments, three of the most important...concerns pertain to air quality, water quality and onshore infrastructure.[4]

Santa Barbarans, meanwhile, complained that the comments of Governor George Deukmejian (who succeeded Jerry Brown in 1983) to Washington on offshore oil leasing did not represent their anti-leasing attitudes. Governor Deukmejian was a conservative who even before his election made clear his intention to abolish the California Coastal Commission. It had more authority and influence over Outer Continental Shelf oil activities than any other state agency, including the governor's office. And it was holding offshore oil projects to stiff standards. Deukmejian slashed the commission's budget severely, but was not able to eliminate the agency which was created by California voters in 1972 and reauthorized by the legislature in 1976.

California's formidable congressional delegation attacked the new five-year plan, and Hodel agreed to discuss it. A proposal was at least tentatively agreed on that would open 150 tracts, mostly off Northern California, in exchange for protecting 6,310 other tracts from leasing for fifteen years. Hodel agreed, then almost immediately renounced the plan. The congressional delegation accused him of backing out of a deal.

Hodel said it should have been called "a proposal, a suggestion, or an offer," not an agreement.[5] In any case, it opened a stormy California relationship with Hodel that lasted throughout the rest of the Reagan administration. The *Los Angeles Times* quoted Hodel as saying he accepted the agreement only because he feared that without it Congress would renew a moratorium on drilling off the California coast. Hodel acknowledged this a few days later in a letter to the editor of the *Times* in which he wrote: ". . . to impose a moratorium at a time when geologists believe there is vast potential off this coast, is to invite a panic reaction in the next oil crisis."[6]

Congress had placed a moratorium on further California offshore leasing in 1984, and renewed it annually for ten years. But each moratorium was for one year only, and in 1986 Interior was working on its 1987-1992 leasing schedule, hoping that the ban would be lifted before then. In February 1986 Hodel said he would withdraw several areas from future California lease sales. He used the familiar device of eliminating areas that had long been set aside as protected—the Channel Islands National Marine Sanctuary, the federal ecological preserve and the federal buffer zone, all in the Santa Barbara Channel. Other areas were deleted on request of the military which used vast expanses of ocean off California for Navy and Air Force exercises. Withdrawing these regions from the sale was regarded as duplicitous because they should never have been considered. The gambit did nothing to enhance Hodel's standing with the California congressional delegation or Pacific shore communities.

Northern California Slighted

Matters got worse when Hodel announced that no hearings on the five-year plan would be held in Central or Northern California where three of the five California sales were scheduled. The first sales would be off Mendocino and Humboldt counties, but the only California hearing would be in Los Angeles, seven hundred to eight hundred miles away. An Interior Department spokesman said hearings were scheduled where they would attract the most people. Not many people lived in the Northern California counties, but within two years Interior would have a new image of this land of scattered inhabitants.

As protests continued, Hodel lost composure. He said his main problem was having to deal with wealthy West Coast environmentalists who he said had stalled exploration "whatever the cost to the United States."[7] In an editorial, the *Los Angeles Times* protested that Hodel's adversaries were not selfish. It might have added that few of them were wealthy. At public hearings, extravagant oil development was supported by federal officials, oil industry executives, and consultants and lawyers collecting two hundred dollars or more an hour. Partisans on the other side were almost entirely volunteers, collecting nothing for their efforts, traveling and eating without expense accounts, and often taking time off from work. They were usually backed by local officials, seldom people of wealth.

After a ten-month confrontation with the oil industry and the Interior Department, Congress late in 1986 postponed all California offshore leasing until after Reagan, who approved the change, would be out of office. The first sale in the 1987-1992 plan would be off Southern California in March 1989, two years later than originally scheduled. The Northern California sale was delayed from December 1987 to February 1989.

With these and other changes, Hodel in February 1987 released another draft of a national five-year plan for 1987-1992. Again, it included five sales off the California coast. And again, Hodel made much of the "withdrawal" of areas which were already in protected status. Almost nobody outside the oil industry was won over. Twenty-six of California's forty-five members of the House sent a letter to Hodel protesting that coastal resources were imperiled by the plan. Republican Senator Pete Wilson agreed, saying that air quality was especially jeopardized. Wilson introduced a bill to transfer Outer Continental Shelf air pollution controls from Interior's Minerals Management Service to the independent Environmental Protection Agency.

MMS controls, far more lax than state or county standards, undermined the efforts of California and its coastal communities to meet federal and state air quality standards. Congresswoman Barbara Boxer introduced a bill to make waters off California a sanctuary free of further federal leasing. The Santa Barbara County Board of Supervisors told Congress that it could not keep up with the onshore needs of offshore development that was already authorized.

More leasing would overwhelm the county's ability to accommodate the industry.

Already under construction or planned off the coast and onshore were sixteen new platforms, two tanker terminals, five oil and gas processing plants, a liquefied natural gas port, oil storage tanks and a network of pipelines. Not all of these were built, but they were all proposed and the county was stuck with potentially fitting them in to one hundred miles of resplendent Pacific coastline.

In an opinion piece published in the *Los Angeles Times*, Hodel explained, unintentionally, the crux of the problem. He pledged to protect the coast, "but it also is my statutory obligation to develop this nation's offshore energy resources in an expeditious and orderly manner...." His department, then, was responsible for the conflicting responsibilities of promoting and policing oil production. In fact, the separate tasks of conducting the lease sales and regulating offshore oil operations, which had been assigned to separate Interior Department agencies, now were combined in one new office, compounding the conflict. The Minerals Management Service was created in 1982 to relieve the Geological Survey of the regulatory burden which too often became political and conflicted with its excellent scientific work. The Minerals Management Service also took over the management and promotion of leasing from the Bureau of Land Management.

Defying enormous opposition, Hodel in April released the final California plan which was virtually unchanged from the one issued in February. Five states, five counties, the city of Santa Barbara, and nine citizens groups sued Interior over the leasing plan. While California's Governor Deukmejian was ambivalent, the state's attorney general, John Van de Kamp, pressed on aggressively. Adding California to the list of plaintiffs, he called Hodel's leasing plan "a sellout of the interests of the people of California for the benefit of private oil companies." The suits, filed in the Court of Appeals in Washington, D.C., asked that the leasing plan be set aside as a threat to air quality, marine mammals, and other coastal resources.

Northern Californians Revolt

Southern Californians, with Santa Barbara as their hub, were the most vocal opponents of offshore leasing throughout most of the twentieth century. But it was an event in Northern California in

1988 that appears to have finally convinced the Interior Department and the White House that something was amiss in their approach.

Minerals Management Service yielded to northern Californians who insisted on a hearing up north to discuss the offer of 1.1 million acres off their coast to the oil industry. "The rugged Mendocino and Humboldt coastline is considered one of the most spectacular areas in the country," wrote Miles Corwin of the *Los Angeles Times*. "Towering redwoods run to the edge of the sea, which is rich in bird and fish habitats. Millions of tourists a year are drawn to the area."

The region was politically conservative and independent, comprised of fishers, farmers, foresters, and small business proprietors. What did *they* have to say about adding offshore oil to their economy? "Bed-and-breakfast inns offered discount prices to anyone who made the trip to testify against oil development," Corwin wrote. "Merchants displayed large anti-oil signs, sold anti-oil T-shirts and offered anti-oil decals and bumper stickers and posters."[8] The installation of as many as twenty-two offshore platforms with all the accompanying onshore industrialization was seen as a threat to fishing, tourism, and the quality of life.

The federal government hired Eagles Hall at Fort Bragg for the February 3 hearing, and that morning the scene outside was alive with anti-oil posters and sea-life pictures. There were speeches, music, coffee and donuts, and political literature. More than 500 jammed into the hall, 1,000 more outside followed the proceedings via loudspeakers, and an additional 200 watched by video hookup in a nearby church, according to the *Los Angeles Times*. The crowd swelled "to somewhere from 3,000 to 5,000 people," wrote sociologists William Freudenburg and Robert Gramling, in a study funded in part by the Minerals Management Service.[9]

During the two-day hearing, more than 750 registered to testify against leasing. "Some of the voices thundered, other whispered, and still others choked with emotion. Guitarists performed songs that they had composed for the occasion; school children sang music and performed plays that they had written as well."[10] For two days the hearing went well past midnight, but even then time ran out before all who wanted to be heard could be accommodated. The Interior officials conducting the hearing "were sitting there scared to death," said Robert Kallman, then an assistant to Secretary Hodel. "All they could hope for was that the meeting would get over with and they could go home."[11]

It was a presidential election year, and offshore oil policies became a campaign issue. Coastal California, at least, was firm on the issue. It wanted a flat-out ban on leasing along the north coast. The California legislature passed a bill to ban state leasing there. Governor Duekmejian vetoed it in August, but there was another way to get the job done. Less than two weeks before the 1988 election, the State Lands Commission voted to prohibit leasing in state waters off Mendocino and Humboldt counties. "It is absolutely essential to send a message to the next Congress and to the next President that these 214 miles are not fair game," said commission chairman Gray Davis. The governor could not veto that administration action. It took only two votes among the three commissioners, but the action clearly enjoyed broad support. Many members of the California congressional delegation had urged the State Lands Commission to create the sanctuary to demonstrate the state's opposition to oil leasing off the northern coastline.

The federal retreat had already begun. In April, the Interior Department indicated it would postpone the Central and Southern California lease sales. In June Vice President Bush, campaigning for the presidency, called for the Northern California sale to be delayed for more study, and the next day Hodel suspended preparation for the sale. President Reagan in October signed a bill that delayed the Northern California sale until October 1989. After six months of administration backtracking on the offshore oil issue, the Texas oilman was elected president of the United States.

Bush Calls a Halt

During the 1988 presidential campaign Bush proposed that the Northern California lease sale be delayed, and early in his term he set aside the entire California five-year plan. This display of restraint followed several studies confirming that leasing decisions had been based on too little information about threats to coastal resources and too little attention to public concerns.

Californians had protested federal offshore leasing for more than twenty years, and state leasing for much longer. Such adversaries were usually regarded by the Interior Department as selfish, uninformed, and a general nuisance. Now, however, social scientists began warning that the hostility was deep, broad, and rational. Even some judges agreed.

Three weeks before Bush took office, the Court of Appeals in Washington, D.C. rejected most of the charges brought by the states, local governments and environmental groups against the 1987-1992 leasing program. But the court upheld the plaintiffs on one important allegation: the Interior Department had skirted the National Environmental Policy Act by slighting the longterm damage that oil operations might inflict on marine life and birds that migrate through coastal areas. These included gray whales, brown pelicans, and other endangered species that ply and fly the California coast. The three-member appeals court required a new impact study on that issue. One of the judges was Ruth Bader Ginsburg who later ascended to the Supreme Court.

With the recent protest at Fort Bragg, the court of appeals ruling, and undoubtedly other influences, President Bush knew that the old ways would no longer work. He must have been aware, for instance, of reports by three federal agencies that indicated the Reagan administration had seriously minimized the potential resource damage of the leasing program, and that it was proceeding with too little information to assure safe and clean operations.

The independent Environmental Protection Agency and the National Oceanographic and Atmospheric Administration had written their reports in 1988, but they were made public only in April 1989 by California members of Congress via the Freedom of Information Act.[12] Another 1988 report by the Fish and Wildlife Service indicated that leasing off Mendocino and Humboldt counties would imperil natural resources far more seriously than the Reagan administration was ready to admit.

(When the report was discovered, also through the Freedom of Information Act, Fish and Wildlife Service officials said it was an "unauthorized" and misleading document, and issued an "official" report that eliminated many of the damaging assessments. Rolf Wallenstrom, director of the Pacific region of the service who raised the environmental concerns, lost his job in January 1989 after he refused a transfer to a new position.)[13]

Bush, then, had stern high-level warnings of the potential damage that could follow irresponsible offshore oil leasing. During his first month in office he postponed indefinitely sale 91 off Southern California, and sale 95 off the state's north coast. He appointed a cabinet-level task force for guidance on future leasing in these areas, and named his new interior secretary, Manuel Lujan Jr., as task force chairman. California's two Senators—Democrat Alan Cranston and Republican Pete Wilson—and seventeen Democratic House members said Lujan was too biased to be chairman.[14] He had asserted that "the answer to our current energy crisis lies just several miles off our shores," but later noted that he was willing to forego leasing where it might create undue environmental damage. He stayed on as task force chairman.

More significant was the appointment of Robert Kallman as executive director. Kallman for three years was an assistant to the secretary of interior, and was now director of the Minerals Management Service. Just as important, however, was the decade he sat on the Santa Barbara County Board of Supervisors when the county struggled with the oil industry over coastal development. Kallman was a conservative who tended to side with the industry, but usually only after strict controls were in place. He understood, while often not sharing, the public protest against coastal industrialization. He could bring a voice to Washington that was usually not heard from industry executives, cabinet-level administrators, or career bureaucrats. Kallman resisted when the Interior Department urged the task force to proceed without hearings and to rely on information already at hand. Kallman said later:

> I said no, this is supposed to be a fresh look at this whole thing. If we don't have these hearings, if we don't get into the states [California and Florida] we're not going to be getting much....We're going to have to use the same old stuff we've always got, and people will say 'that's passe; that took place two or three years ago.' I just went ahead and did it; I didn't ask anyone's permission.[15]

The task force sat through five California hearings on leasing plans for the state's northern and southern coasts. These, plus a Florida sale, had been delayed until President Bush heard from the task force after the hearings. As expected, the task force got an earful. According to the task force report:

> In its nine public workshops...the task force heard a wide variety of concerns expressed about undesirable environmental effects that people believe could result from OCS oil and gas activities. . .In part because of the oil spill from the *Exxon*

> *Valdez*, the risks of oil spills from offshore oil development were a foremost concern. Other concerns included air quality, conflicts with fishing, effects on tourism and recreation, changes in onshore land use and 'lifestyles,' and effects on sensitive environmental areas and ecosystems.[16]
>
> ...the vast majority of elected officials, coastal dwellers, and marine resource users who attended the workshop were vocally opposed to any more oil and gas development on the OCS....Commercial fishing, tourism, water quality, environmental regulation, and sensitive marine habitats were the topics of the day.[17]

Kallman said the president did not interfere with the work of the task force and that Lujan appeared pleased to have Kallman initiate policy proposals. The task force recommendations, influenced and reinforced by findings of a National Research Council committee, came in a series of options, all of them in conflict with earlier Interior Department leasing procedures. For all three sales, the task force recommended cancelling, postponing, and/or reducing the areas offered. It urged tighter standards for oil spill contingency plans, and air quality requirements as stringent as those enforced onshore by state and local governments.

What's a Lease Worth?

A nagging problem since the beginning of federal leasing was the lack of information about the amount of oil in the lease sale areas. Oil companies could not drill exploratory wells where they did not have leases. They had only core samples and geophysical survey data that only indicated geological formations that *might* contain oil. As oil officials were wont to say, you never know until you drill. As little as the oil industry knew before the lease sales, the federal government—which was auctioning the tracts—knew even less. William Pecora, then director of the Geological Survey, testified in March 1969 that the federal government was "flying partly blind" in offshore lease sales because it did not know the value of the tracts up for bids.[18] Later that same year Interior Secretary Walter Hickel suggested that federal agencies should "get their own geological information. That was not too well accepted," he said later. "I'm not real sure that the idea wasn't good."[19]

The president's task force and the National Research Council committee were concerned about the government's lack of information. The government was uncertain of the value of leases it was awarding, and coastal communities had no idea what to prepare for—dry wells with almost no onshore impact, or massive industrial complexes that overwhelmed local planning efforts. Consequently, it was impossible to prepare helpful impact reports for lease sales. It had to be done after exploratory drilling was completed. This created two fundamental problems, according to the National Research Council/National Academy of Sciences report:

> First, the exact location of oil and gas reservoirs is unknown at the pre-lease stage. As a result, it is impossible to identify the specific future location of facilities and to predict specific environmental impacts of development. Equally important, the uncertainty about actual oil and gas reserves at the pre-lease stage makes it difficult to balance the national benefits of production against the environmental risks.
>
> Second, by the time producing reservoirs are identified, the industrial lessor typically has committed enormous amounts of money to the lease....[A] decision to lease is generally perceived as tantamount to a decision to develop and produce....[O]nce it does become possible to generate the needed information and analysis, a decision not to proceed with development has already been effectively precluded....
>
> If a more comprehensive assessment for the entire possible sequence of OCS oil and gas activities cannot be made at the time of the leasing decision, then leasing and exploration should be more clearly separated in practice from development and production, with distinctly different phases of scientific data-gathering and analysis.[20]

The Interior Department did not support the idea in 1969, and it did not appear to accept it in 1990.

Efficiency, Conservation

The president's task force took much public protest seriously, but was not impressed with demands for federal initiatives to reduce oil demand through efficiency and conservation. The group echoed the approach of the oil industry: meet demand primarily via increased domestic production. Amory and Hunter Lovins, noted energy economists, disagreed. In January 1990, they wrote:

> For nine years through 1985, the United States was making such rapid progress in saving oil that, if we'd just kept up the same pace, we'd have needed no Persian Gulf oil from 1986 onward. But instead, the Reagan Administration's 1986 rollback of light-vehicle efficiency standards immediately doubled oil imports from the gulf: It wasted oil as fast as the govern-

ment hopes to extract oil from beneath the Arctic National Wildlife Refuge....Our [Persian] Gulf dependence is no accident. It was deliberately caused by dimwitted federal energy policies that persist to this day.[21]

The Lovinses were noted for proposing outrageous energy-efficiency policies, and later being proven to be too conservative in their projections. While some utility companies slowly adopted their recommendations, the federal government dug in its heels.The president's task force, for instance, acknowledged that much public protest stemmed from the lack of federal support of energy conservation and efficiency and alternatives to petroleum fuels. But the panel was not yet ready to give these appeals much credence.

If the Bush administration was not prepared to end the petroleum era, it was also not eager to continue the offshore California leasing binge. In June 1990, one month after the task force report was leaked to the public, President Bush cancelled the three scheduled California lease sales. He said there would be no more sales off California until the turn of the century, except—alas—in the Santa Maria Basin off Santa Barbara and San Luis Obispo counties, and in the Santa Barbara Channel. These two petro-battlefields would be spared only until 1996. They would be part of the 1992-1997 five-year plan which was already being drafted.

Farther north, Bush approved NOAA's recommendation to establish a 2,200 square-mile marine mammal habitat and breeding grounds as the Monterey Bay National Marine Sanctuary. That banned oil leasing there. The president also cancelled a scheduled sale off Washington and Oregon. Sales in the north Atlantic and eastern Gulf of Mexico were also scrubbed.

The California Coastal Commission, in a unanimous vote, and Pete Wilson, now California's governor, protested that Santa Barbara Channel and the Santa Maria Basin should also be exempted from oil leasing until after the turn of the century. The stretch from Oxnard to Morro Bay includes coastal areas of "extreme environmental sensitivity" which should be spared the drill bit, Wilson wrote in a memo to the Interior Department. The governor's office held a public hearing on the new five year plan in Santa Barbara in September 1991, and heard only opposition.

Why was Santa Barbara County again a leasing target? Kallman said he asked a colleague in the administration. "He said the President was convinced that as long as there was already so much activity there that a little more activity is not going to make much difference," Kallman said.

"I said, `Don't you realize that there are almost 90 leases there that haven't even been explored, much less developed, and that's enough to keep them busy for 30 years? Why do we need more of these?'"

Before Bush left office he received another report from the National Research Council's panel which was assigned by the Interior Department to study socioeconomic consequences of offshore leasing. Within its 153 pages, the report described, sometimes poignantly, the emotional impact of being told of plans to industrialize one's waterfront. Public attitudes about things that cannot be demonstrated in numbers must be seriously considered, the report noted.

> [A]pprehensions about esthetically unpleasing developments, for instance, or about decreased recreational possibilities, environmental pollution, or other diminishments of life's quality—and not simply the actual developments themselves—are real and immediate effects, and they are likely to lead to further ones....Moreover, uncertainty concerning the future of a region ineluctably increases from the moment a tract appears on a five-year plan until the failure of a lease sale or until exploration ends in either abandonment of the leasehold or in drilling. Such uncertainty also constitutes a real impact of OCS activity....
>
> [T]he psychic and social tensions that attend uncertainty, or anger at and alienation from the government, cannot be represented adequately, or even at all, in quantitative terms of any sort, let alone monetary terms.

Attempts to mitigate violations of philosophical or esthetic values through cash awards were likely to be taken as insults heaped on previous injuries, the authors said. It posed the question, How much is your integrity worth?

Basic values such as liberty or happiness, onto which one cannot tie a price tag, "tend, in their nature, to be very low in specificity," the report continued in a mood uncharacteristic of oil leasing memoranda to the Interior Department:

> To say, however, that values are not specific or even vague does not say that they are not cogent, or even decisive, in the formation of positions on which social actors stand, from which they define the general conceptual, social and geographic territory which they feel is rightfully theirs, and from which they will act. It could well be that there is a direct correlation between the vagueness of a value and the strength

> of the motivations it engenders. People will put their lives in danger to protect whatever they mean by 'liberty' or 'democracy,' but not to balance the budget.

If only Stewart Udall had had such a report on his desk in 1968.

"There is a tendency for agencies (and many other specialists) to dismiss public concerns as not being expert," the report stated. "But *everybody* is an expert on her or his fears, desires, wants, needs, and values. And this is crucial when describing an approach to obtaining socioeconomic information."[22]

With a stack of reports on his desk condemning the management of the Reagan-Watt-Hodel Pacific oil rush, President Bush, conservative Texas oilman, suspended the California leasing program.

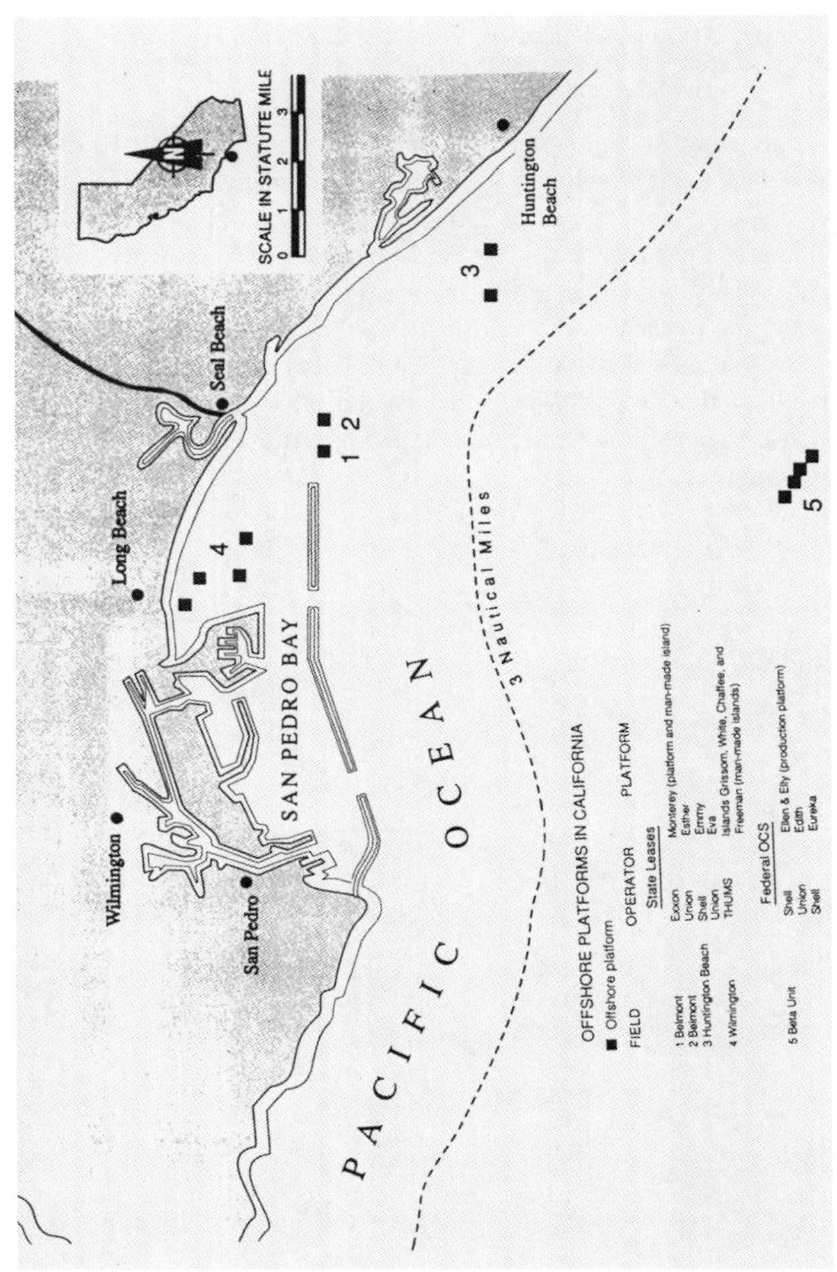

Coastal Area Off Los Angeles and Orange Counties

Chapter Ten

State Leasing Revival Fails

The heavy bidding for federal offshore leases between Point Conception and Point Arguello in 1979 made lights flash in Sacramento. "If there is that much oil out there, the state wants to make sure it gets its share," said Robert Kuehn of the State Lands Commission staff.[1] Consequently, he said, the commission would lease state waters between the shoreline and the federal leases. The state had not awarded major offshore leases since the 1960s, and never north of the Santa Barbara Channel. The news stunned Californians who had routinely relied on the State Lands Commission for support against federal leasing.

The state's suspicions in 1979 about the abundance of oil off the southwest corner of Santa Barbara County were fortified by enormously high bidding there in another federal sale in 1981. So the state planned a sale, covering forty thousand acres, for August 1983. At its first meeting of that year, the Santa Barbara County Board of Supervisors argued it would go to court to ban or delay the sale. Too much was at stake, the supervisors said, in a region of great species diversity and where too little was known about the marine life to subject it to the encroachment of the oil industry.

On May 25, 1983, the California Coastal Commission rejected the lease sale on a twelve to zero vote. Even the four appointees of Governor Deukmejian voted against the state leasing plan. But the Lands Commission maintained it did not need a Coastal Commission permit, and would proceed with the sale. The governor's office tried to get the two commissions on the same track. But both were largely independent of the governor's office, and the effort was almost useless.[2] The Santa Barbara Environmental Defense Center sued for a determination that the sale was a "development," and as such required a Coastal Commission permit. The State Lands Commission insisted that offshore leasing was not a "development."

Faced by two lawsuits, however, the Lands Commission postponed the sale six weeks to September 30, 1983. In the meantime, the Coastal Commission on August 23 reversed itself and voted six to four to approve the sale despite hearing several hours of protest. A Coastal Commission majority was placated by Lands Commission concessions, including a ban on oil activity within a half mile of shore and in waters less then fifteen fathoms (ninety feet) deep. No marine tanker terminals would be allowed, and oil activities would be banned for six months of each year in halibut trawling grounds. Protesting that these changes were not nearly enough to protect the region's rich resources were Santa Barbara city and county, State Senator Gary Hart, State Assemblyman Jack O'Connell, environmental groups, and fishers.

Ecological protests failed, but environmentalists had a procedural objection. They received copies of the commission staff report only the day before the Coastal Commission hearing. State law, they said, required more time to review such documents. With the lease sale date a week away, Superior Court Judge Patrick McMahon set aside the Coastal Commission's approval, ruling that the short time available to review the staff report violated the state administrative code. He told the Coastal Commission to conduct another hearing, and told the Lands Commission not to open any lease bids in the meantime. That was not the last freakish twist in the serio-comedy.

Judge McMahon had nullified the Coastal Commission's approval of the lease sale, but now he was faced with allegations that the Lands Commission had acted illegally in calling for the sale in the first place. There were four Lands Commission hearings on the proposed sale in 1982, but there was no agreement on which hearing the decision was made to approve the sale. That was important because a sixty-day notice of such a hearing was required, and no approval was effective until thirty days after the decision.

The Lands Commission had delayed the sale again to November 1983 and, now as complications mounted, postponed it for the third time, this time indefinitely. Still insisting it did not need Coastal Commission approval, the Lands Commission boycotted the October 26 Coastal Commission's third meeting on the issue. Environmental and planning groups urged the commission to withhold approval, asserting there was too little information about the threat to coastal and marine resources, too little resource protection required, and too much to cope with on top of oil activity already in progress. After rejecting the sale at its first hearing, and approving it at the second, the Coastal Commission reversed itself again and voted against it. There the issue dangled, unresolved, throughout the 1980s. The Lands Commission neither conceded any ground to the Coastal Commission, nor rescheduled the lease sale.

It shifted its attention instead from Santa Barbara County to the northern counties of Mendocino and Humboldt where in October 1988 it created an offshore state sanctuary in response to the federal government's plan for OCS leasing. So while the Lands Commission promoted leasing in one area, it banned it in another. The seemingly contradictory policies were in fact quite consistent. Federal leases already existed off Santa Barbara County, and the state wanted to prevent drainage of its oil to adjacent federal leases where production was already authorized. No federal leases were yet sold off Mendocino and Humboldt counties, and the state wanted to keep it that way. State leasing there could have encouraged federal action to avoid drainage losses. Creating a state sanctuary there was meant to advise Congress and the Interior Department that the state was serious about preventing Northern California leasing. The legislature twice passed bills that would have strengthened the sanctuary by placing it under statutory law rather than simple administrative decree, but Governor Deukmejian vetoed both bills.

In a limited and indirect way, the state was protected from drainage losses from production in adjacent federal waters. Robert Hight, executive officer of the State Lands Commission, explained:

> The state does receive 27 percent of all bonuses, rents and royalties from federal leases lying wholly or partially within the first three miles of the federal Outer Continental Shelf....All coastal states with active federal leases within three miles of state waters are eligible to receive these funds as compensation for onshore impacts of the federal lease operations and as compensation for any enhancement that state leases have made to the value of the federal leases.[3]

In December 1989, the Lands Commission made sanctuaries of all state waters that were not already leased. It was an administrative move that could be reversed by any two Lands Commission members, and parts of the sanctuaries were only temporary. For the moment, however, the sanctuaries were safeguarded by Lieutenant

Governor Leo McCarthy and State Controller Gray Davis who constituted an anti-oil majority on the commission.

Get Out Oil hired a small boat to register its protest off the "Summer White House" at San Clemente during the Nixon Administration (*San Clemente Sun-Post* photo)

Chapter Eleven

After Leasing, The Hard Part

The federal offshore oil leasing binge ended in 1984, but that was only the beginning. In eight major sales since 1967, the Interior Department had awarded 311 leases covering more than 1.6 million acres off the state's coast. Production continued in the region of the 1969 blowouts, and two immense new fields and other lesser oil deposits had been discovered off Santa Barbara County. It was time now to install the massive industrial structures to produce, process, store and transport the oil. ARCO compounded the coastal dilemma with a plan to add three platforms in state waters just off the university campus west of Santa Barbara. If federal and state agencies issued offshore permits, the county was faced with land use planning, air pollution control and providing public services.

As the demands threatened to overwhelm the county, new governmental structures were needed to process permits and oversee development. At the same time, however, the Board of Supervisors, which frequently had a conservative majority, tried to "streamline" the permitting process. The two efforts often conflicted. The planning system had burgeoned awkwardly as the county struggled to comply with the environmental legislation of the early 1970s. While the system needed streamlining, which usually meant truncating, more people were required to contend with mounting oil industry demands.

In 1973 the supervisors created an Office of Environmental Quality and appointed Albert Reynolds as its director. Reynolds came from San Diego County where he had been executive assistant to the director of the county's Environmental Development Agency. He was innovative, daring, honest, and sensitive to the value of natural resources. His office was elevated to a Department of Environmental Resources in 1977. In 1981 Reynolds was instructed to merge his agency and the Planning Department. The supervisors approved his plan for a new Department of Resource Management, and made him interim director while it pondered the appointment of a permanent director. Reynolds was widely expected to get the job.

Instead, the board on a three to two vote hired Dianne Guzman, assistant planning director for San Diego County. Reynold's backers felt the move was not so much hiring Guzman as it was firing Reynolds. The conservative majority of county supervisors felt he had been too firm with the oil industry and other developers. Reynolds, however, insisted he had effectively brought the oil industry and county agencies to the table without the hostility that was perceived from the outside.

Guzman was a professional planner and beyond that, a skilled politician. She had to retain her one-vote majority support among the county supervisors, deal with the demands of a strong environmental community, and keep an enormous offshore oil industry from industrializing the shoreline. While the supervisors hoped to shrink the planning staff, Guzman in 1982 induced them to expand her department by adding a division to deal with oil issues. The county also established an Air Pollution Control District which quickly became a corps of experts on complex air quality problems, most of which stemmed from the offshore oil industry.

By 1983 the county faced a task that would confound most states—or most nations, for that matter. It was trying to accommodate nine large oil industrial projects simultaneously. More were added in 1984. Planning for any one of them was a challenging responsibility. A plan encompassing all the proposals was required, and the county made efforts in that direction. The oil companies did not cooperate among themselves when it came to comprehensive planning. Some legal barriers also frustrated integrated planning. State law, for instance, gave counties one year to rule on a project application, whether for a residential lot split or a multi-million industrial oil complex. Time for processing permits could be extended only if the applicants consented. Coordinating planning was impossible because each industrial plan required a separate impact report and all were on different schedules. Some applications were submitted in fragments, precluding a review of a single project in its entirety, much less coordinated with other plans.

Because of this fragmented approach, the County Planning Commission and the Citizens Planning Association maintained that Santa Barbarans were not getting "the big picture," the full scope of the oil industry's plans. This despite repeated references to estimates that by the early 1990s, production off the county's shores would amount to 500,000 barrels a day. Even though the forecast was overblown, Santa Barbara Channel oil production was approaching 200,000 barrels a day in the early 1990s. And it was far from its peak.

Planning for onshore construction was primarily a county role, but the projects crossed jurisdictional lines, requiring state and federal permits. Each level of government defended its turf and pursued its own interests. Strife was compounded by conflicts between agencies within each level of government. County agencies usually worked in concert, supported by the city of Santa Barbara. But on the state level, as we have seen, the Lands Commission and the Coastal Commission sometimes fell into civil war. In Washington, the Environmental Protection Agency and the Minerals Management Service for thirteen years vied for air pollution authority on the Outer Continental Shelf. Even agencies within the Interior Department often were in conflict.

Largely on the initiative of Dianne Guzman, director of the county's Resource Management Department, joint review panels, representing all jurisdictions, were organized to work together on permit applications. But the projects continued to be reviewed one-by-one rather than as an integrated planning endeavor.

Exxon, in Deep Water

After a bitter and protracted hassle, Exxon installed Platform Hondo in the Santa Ynez unit which comprised seventeen leases totaling nearly one hundred thousand acres in the middle of the Santa Barbara Channel. (See Chapter Six) One platform was not enough to exploit the riches of the unit which was described by the *Oil and Gas Journal* as having the greatest potential among American offshore fields, and fourth largest in the nation—offshore or onshore. The trade magazine noted the field had a production potential of one billion barrels of oil, an estimate that some "experts" said was conservative. The "field" was in fact a combination of four fields within the Santa Ynez unit. Organizing several leases into production units was required by the Interior Department to reduce the number of platforms necessary for production. The number was also reduced by having one company operate the unit for all the companies with financial interests in the leases. Exxon was the operator for the Santa Ynez unit.

The 1969 catastrophe was dramatized by massive oil contamination. Now the concern had shifted to air pollution. In 1977, Platform Hondo was in place and soon would be producing. But it was not yet determined whether Exxon would process its oil onshore under strict county air quality controls, or in a floating plant in federal waters where regulations were less stringent.

For ten more years, Exxon would threaten to process its oil offshore if the county insisted on air quality controls that it maintained were necessary to meet federal Clean Air Act requirements.The offshore arrangement meant that oil would go through an Offshore Storage and Treatment (OS&T) processing plant and then transferred to tankers, notorious air-polluters. Testing its authority, the Environmental Protection Agency in 1977 indicated it would enforce its standards on the offshore plant.[1] But the Interior Department claimed that authority exclusively under terms of the Outer Continental Shelf Lands Act. Its rules were much more attractive to the oil industry.

The Interior Department prevailed in federal district court in 1978 and on appeal in 1979. That resolved one dispute but left another. One federal agency (EPA) was telling Santa Barbara County to meet Clean Air Act standards, and another federal agency (MMS) was undermining county efforts to meet those standards. The Interior Department asserted that not enough air pollution from the OCS reached onshore to make a significant difference. To its credit, Exxon installed vapor-control technology on its offshore processing vessel to greatly reduce the volume of escaping fumes. But the Environmental Protection Agency, the county, and state agencies protested that pollution from offshore oil operations would still overwhelm their efforts to clean up the air onshore.

To offset air pollution wafting in from offshore, the county established strict air quality standards on anything Exxon planned to build onshore. Supported by the Interior Department, Exxon protested that this was an unlawful attempt to extend county authority onto the Outer Continental Shelf. Exxon made good on what many thought was a bluff, and installed the processing plant offshore. Production from Hondo started in 1981. The next year Exxon submitted plans for three more platforms in the Santa Ynez unit, and kept its options open to build an onshore processing plant with a capacity of 140,000 barrels a day, or to enlarge its offshore plant to handle 80,000 barrels. To maintain its freedom from California air quality controls, it would have to produce at far less than capacity.

Meanwhile, natural gas from Hondo went to a new plant completed in 1983 by Pacific Offshore Pipeline Company (POPCO) in Las Flores Canyon west of Santa Barbara The gas market was just a few pipeline miles away.

The California Coastal Commission in June 1983 agreed to the platforms, but delayed acting on expanding the floating processing plant because it was still under environmental review. Exxon and the county spent much of 1983-1984 arguing whether the company would transport its increased production by tanker or pipeline. The county and Exxon agreed in June 1984 on temporary tankering until a pipeline system became "economically feasible." For ten years, Exxon had refused to agree to let any public agency define feasibility. Exxon's conceding that determination now to the county was a breakthrough.

But the county was no closer to ending the air pollution threat. The Air Pollution Control District maintained that the Exxon project would violate air quality standards "all the way to Carpinteria and Ojai," approximately fifty miles east of the platforms. The APCD warned of pollution from tanker loading, drilling machinery, oil and gas processing, and leaking storage tanks. When the County Planning Commission met in July 1984 to consider Exxon's plans, opponents packed the twelve-hour hearing. When the hearing resumed a few days later, the commissioners voted three to two against a tanker terminal off Las Flores Canyon. The commissioners feared it could become an air pollution hot spot, and said another location should be considered.

Exxon resorted to its familiar threat to process its expanded production offshore beyond the county's jurisdictional reach. "Exxon is behaving like a little boy taking his marbles and going home, except this is a little more serious," said Commissioner Joan Wells. "Sometimes we have to stand up and draw a line," she said, then added for assurance: "They'll get their terminal." The County Planning Commission approved the onshore processing plant in August, largely to induce Exxon to bring its oil onshore for processing.

The Board of Supervisors supported the Planning Commission, assuring Exxon it would receive a tanker terminal permit after more information became available. The question was location, not whether there would be a terminal. But Donald Cornett, project manager for Exxon, insisted that "you leave Exxon with no alternative but to use its OS&T [Offshore Storage and Treatment] facility."

Exxon then asked Secretary of Commerce Malcolm Baldrige to approve expanding the OS&T vessel despite the objection of the state Coastal Commission. Baldrige preferred to see the issue worked out locally, and the dispute continued during three years of hearings, conferences, close-door deliberations, trips to Washington, compromises, lawsuits, mutual distrust, and agreements that did not survive.

Agreement came close in the summer of 1986. On a three to two vote, the Board of Supervisors dropped several air quality requirements that had been recommended by the County Planning Commission and the energy planning staff. Instead, it reserved the right to impose more restrictions later if the original regulations were insufficient.The struggle that had consumed much of the county's time and emotions for four years appeared to be resolved. But no. The next week Supervisor David Yager changed his mind and voted for the tougher up-front air pollution controls. Incensed, Cornett accused Yager of caving in to environmentalists' pressure. Yager replied that if Cornett believed that radical political groups controlled county politics, "in this case, the kooks are right." Resorting to Exxon's familiar weapon, Cornett said the company would not accept the supervisor's terms, and would ask Commerce Secretary Baldrige (again) to allow expanded use of the OS&T. "You can stick by your agreement," Cornett told the supervisors, "or you can stick it in your ear."

(It was described by one reporter as Cornett's "swan song," as he was being reassigned to Exxon's operations at Valdez, Alaska. He would have ample opportunity there before long to coin more flamboyant phrases for far-ranging publication).

For the third time, Exxon had won a permit it would not accept. It declined a California Coastal Commission permit in 1976 because it did not like a pipeline condition. It declined a county permit later because it did not include a marine terminal approval which the county said would come later. And now it rejected a county permit because of stringent air pollution controls. It became known as the company that would not take yes for an answer.

Most striking, however, was its continued ability to convince much of the press that the permits had in fact been *denied*. As we have seen (in Chapter Six), Exxon conned much of the press into reporting, or strongly implying, that the California Coastal Commission had denied a permit for Exxon's first phase of the Santa Ynez unit development. Now the *Santa Maria Times* reported that the supervisors had "in effect" forced the company to move its project offshore. In an editorial, the newspaper reported Supervisor Toru Miyoshi of Santa Maria had "voted against Exxon," while in fact he voted for the permit which Exxon rejected. The newspaper carefully avoided reporting that the permit had been approved.

At least one Santa Barbara radio station reported that the permit had been denied. Even the *New York Times* reported that ". . .the board of Supervisors, after one member, David Yager, reversed an earlier vote, withdrew approval of the permit."[2] County delegations travelled to Santa Maria to clarify what the supervisors had done, and to Washington to advise the Interior and Commerce departments of what really occurred.

As usual, Interior was impatient with state and local insistence on clean and safe coastal oil operations. It had expressed irritation with California Coastal Commission regulations in 1976, and now ten years later it railed against the county for its treatment of Exxon. William Grant, who was in charge of the Minerals Management Service on the West Coast, wrote to the *Santa Barbara News-Press*:

> On September 3, Santa Barbara County imposed new conditions to control air emissions from Outer Continental Shelf platforms. We believe these conditions are unreasonable and exceed the county's legal authority....MMS must conclude that the offshore option is the only available way to develop the Santa Ynez Unit's considerable reserves.[3]

The letter illustrated again the Interior Department's dual and conflicting role of promoting and policing the offshore oil industry. The county hired two "insider" lobbyists: Edward J. Rollins, former political tactician for President Reagan; and William Ruckelshaus, twice EPA administrator and once deputy attorney general. Rollins, a conservative political consultant, was hired "mainly for access" to the White House, Supervisor Bill Wallace said. "He was successful although I don't think he was philosophically on our side."[4]

The county coaxed Exxon back to negotiating. The strategy this time was to get all interested parties together on an informal agreement before revisiting another impasse in public hearings. Environmentalists were part of the non-official process. By mid-1987 the group had crafted a plan that was expected to reduce air pollution from the project by sixty-five percent. Installation of one of the three platforms would be delayed, power on the other two would be electric instead of internal combustion, exhaust controls would be added to crew and supply boats, and the crude would be shipped via pipeline, not in tankers. Before the end of the year the revised approach was approved unanimously by the Board of Supervisors and in February 1988, by the California Coastal Commission.

The county had hung tough and won most of what it sought, but it took much of the 1970s to approve the initial phase of the development, and much of the 1980s for the second phase. That was the effort required to accommodate only one company's endeavors. Other oil corporations were pressing their massive programs at the same time.

The Point Arguello Project

Chevron and its seventeen partners applied for permits in 1983 for the Point Arguello project which was said to rival or exceed the Exxon project in production potential and in its assault on county financial and natural resources. Chevron and the Minerals Management Service described the field off the southwest corner of Santa Barbara County as the biggest domestic find since the 1968 Prudhoe Bay discovery in Alaska. The impact report for the project noted there were three billion barrels of oil in place, of which 300 million to 450 million barrels were recoverable.

The Arguello project included three platforms ten to fifteen miles offshore and a large oil and gas processing plant at Gaviota between Santa Barbara and Point Conception. The platforms and the plant would be linked by twenty-five-mile oil and gas pipelines. Tankers would take on oil at a waterfront terminal and would moor at an offshore buoy linked by an underwater pipeline to onshore tanks.

When hearings opened in 1984, residents lined up for hours to protest pipeline hazards, interference with commercial fishing, oil tanker traffic, air pollution, desecration of Chumash archeological sites, interference with cattle grazing, the relocation of a historic school, truck transport of petroleum byproducts, interference with coastal resources, and environmental threats to twenty-six streams, two wetlands, and nineteen other sensitive habitats.

After attaching stringent conditions, the county and the California Coastal Commission approved a permit to build the Gaviota plant. The commission staff explained that the project violated at least six environmental policies—or would have without the California Coastal Act exceptions for offshore oil projects. These loopholes were believed necessary in 1976 to get the Coastal Act through the legislature.

After clearing all political, legal, and administrative obstacles, Chevron broke ground with confidence. But when earth-movers went to work in February 1985 they scooped up heavy concentrations of polychlorinated biphenyl (PCBs). Before Congress banned PCBs in 1979, they were used in a gas processing plant long since abandoned and demolished. They were in heat-transfer fluids that leaked from pipes and contaminated 3,900 cubic yards of soil. Concentrations of 50 parts per million were believed to be enough to cause cancer. Concentrations here, a few hundred yards from a schoolhouse, were recorded up to 3,700 parts per million, although the average over most of the site was "only" between 50 and 370 parts per million.

Chevron had not spilled the toxic chemical; the protests centered on attempts to cover up the dirty discovery, or at least failing to notify the appropriate agencies. Chevron knew about it in February 1985, and informed the Environmental Protection Agency which did not tell anybody in the county about it. The county had an "onsite environmental coordinator" at the scene who knew about the PCBs, but there was nothing in his instructions about that, and he did not report it.

The principal of Vista Del Mar school, three hundred yards from the dusty, wind-blown grading site, became curious when she saw men in white "space suits" removing soil. Little by little, the word seeped out until county supervisors one year later complained that they were the last to know. Construction was delayed until a mountain of contaminated dirt was hauled to a dump in neighboring Kern County.

By late 1988 the plant was completed. Oil and gas pipelines were in place, running nineteen miles five feet underground across two ranches and Gaviota Beach State Park. Hollister Ranch, where more than 800 people lived on 100-acre parcels, felt threatened by the gas pipeline. Toxic and explosive hydrogen sulfide up to 7,000 parts per million (ppm) would flow with the natural gas. After long, contentious hearings, the county and state decided the pipeline was safe. But again, there were unsettling discoveries. Before startup, it was disclosed that the natural gas in the Arguello Field was contaminated with up to 22,000 ppm of hydrogen sulfide, more than three times the original estimate. Inhalation at 1,000 ppm is fatal. Everything stopped for more hearings.

They dragged on through 1988 and three months into 1989. Much of the scientific and technical testimony was over the heads of the audiences, but the hearings were packed. People knew hydrogen sulfide was deadly, that the concentrations would be three times greater than described initially; and that accidents happen. Chevron made its case in a half-page newspaper advertisement; opponents responded with full-page ads.

After Chevron agreed to substantial changes, the Board of Supervisors in April 1989 voted four to one to approve the pipeline with the higher toxic level. The company agreed to reduce gas pressure in the line by thirty-three percent, and bury sensitive leak detectors just above the underground pipeline. After the gas was purged of hydrogen sulfide at the Gaviota plant, Southern California Gas made it available for sale to its customers.

By mid-1991 the Chevron partners were finally producing oil, but at less than capacity. The dispute over whether tankering would be allowed and under what conditions occupied the first half of the 1990s, a conflict discussed in a subsequent chapter.

ARCO Closes In

Although the Exxon and Chevron industrial sites were rather remote from urban areas, the public protest was intense. But the Atlantic Richfield Company created an even greater outcry when it sought permission to establish three double-unit offshore platforms within two miles of UC Santa Barbara and the student community of Isla Vista.

ARCO had installed Platform Holly in the South Ellwood field in 1966. But the true potential of the field was not discovered until later, and in 1983 ARCO unveiled plans for a dramatic array of platforms to produce the new-found oil. In each of three new locations there would be two platforms linked by a steel bridge, or causeway. They would push ARCO's daily production from 10,000 barrels to 80,000 barrels.

This was not what the University of California had in mind when it moved the Santa Barbara campus to the oceanfront in 1954. The tidelands sanctuary, which prohibited oil leasing, extended westward to Goleta Point, not far enough to protect the waterfront of the campus and future student community.

Chancellor Robert Huttenback responded early and firmly. "It is now apparent the Santa Barbara Channel has become the location for the largest offshore industrial development in the Western Hemisphere," he protested. And that development was now approaching within two miles of the campus's vaunted shoreline. Daniel Morse, director of the UCSB Marine Science Institute, said that the ARCO project could close down the institute. "Marine science is one of the largest and internationally recognized areas of research and teaching at UCSB," Morse said, and it depended on a constant 40,000 gallons per day of non-polluted seawater drawn from the ocean near the campus.

More than fifty UCSB courses made use of the offshore resources, and much advanced research on marine organisms depended on the seawater lab. It was not just oil spills that troubled the university. Dumping wastewater, drilling muds and drill cuttings would create turbidity, turbulence, ocean-bottom disturbance, contamination and probably toxicity. Noise, air pollution and the expansion of nearby processing plants also threatened to close in on the university community.

Because the ARCO leases were in state waters, the State Lands Commission would decide the fate of the project. During more than two years of hearings, the university was joined in protest by the Santa Barbara County Board of Supervisors, commercial fishers, Isla Vista residents, and scores of Santa Barbarans who protested the intrusion into their ocean view and recreational waterfront.

ARCO had unwittingly mobilized the university community. Never before, as far as anyone could remember, did a controversy so solidly unite the administration, staff, faculty, and students. During the dispute, Huttenback was succeeded by Daniel G. Aldrich Jr., who as interim chancellor picked up the spear and led the university charge.

The county and the university wanted ARCO to forego Platform Heron, which would be just off the campus and Isla Vista. ARCO argued that was the project's kingpin, and had to stay. Given ARCO's firm stand on that issue, and the overwhelming public protest, the State Lands Commission on May 27, 1987 denied the project, two to one. Lieutenant Governor Leo McCarthy and State Controller Gray Davis, Democrats with plans for higher office, supported what they saw as overwhelming public sentiment. The third vote was held by state Finance Director Jesse R. Huff, appointed by Governor George Deukmejian. Huff was not on hand for the preordained decision, sending instead a deputy to cast the minority vote.

Richard Ranger, representing ARCO, complained that the Lands Commission staff's final report relied too heavily on aesthetic considerations in recommending denial. The report noted that the visual insult on the seascape was the most nagging problem among those who testified. "Platform Heron is of such a scale that the mind cannot readily block it out from view," the report said. "Concerns about the visual impacts of....Heron were expressed from all quarters of the community."

Nine days before the ARCO vote, Interior Secretary Donald Hodel came to Santa Barbara to promote his federal offshore leasing program. In an attempt to placate Santa Barbarans, he held up a dime between his thumb and index finger. "This is how big an offshore platform appears from onshore," he said.

Here were the oil industry and the Interior Department telling coastal dwellers familiar with the sight of offshore oil platforms that what they were seeing should not offend them. But such a view *did* offend them. Five years later the National Research Council demonstrated that such public concerns are real and deeply felt, and advised public agencies not to slough them off as "merely" emotional.[5] (See Chapter Nine).

After the vote, ARCO sued the state, alleging that it had no authority to prevent full development of state offshore leases. But the industry had argued in 1981 that there was plenty of opportunity after a *federal* lease sale to impose environmental safeguards. (See Chapter Eight.) Now ARCO was arguing that a *state* lease conveyed the right to produce oil without serious government interference. ARCO contended that if it could not proceed with its plan, the state must buy back the leases for $793 million. The suit dragged on into the 1990s when a settlement was reached but ran into complications described in Chapter Fifteen.

Union Oil, Again

Exxon, Chevron and ARCO taxed the patience, resources, and stamina of Santa Barbara County for years before their projects were approved or denied. The companies, of course, felt that the county had inflicted undue demands. But Union Oil, which suffered the wrath of the living world in 1969, submitted a plan in 1982 that—relatively speaking—almost slithered through the permit process. The company, which became Unocal in the meantime, wanted an offshore platform just north of the Point Arguello field off Point Pedernales in the Santa Maria Basin. Although the impact report described severe assaults on natural resources, the proposal for several reasons drew relatively slight opposition. Only one platform was requested. It would be off Vandenberg Air Force Base, a non-urbanized region. The processing plant would be several miles inland in the north county where the oil industry enjoyed friendly relations. No tankering would be involved; oil would go from the platform to the processing plant and then to Union's refinery in San Luis Obispo County, all via pipeline.

POPCO Scare

Oil companies are usually also gas companies because oil and gas ordinarily are found in the same geologic formations and they come up together. Each is processed separately. An enormous gas processing complex approved during the 1980s was built by Pacific Offshore Pipeline Company (POPCO), a subsidiary of Southern California Gas Company. It was built in Las Flores Canyon twenty miles west of Santa Barbara, near the coast but out of sight from the beach or U. S. 101, the coast highway. Thirty million cubic feet of natural gas would flow daily from Exxon's offshore Santa Ynez unit, and the gas plant was designed to double that capacity with additional units to be built later. With the plant nearly complete, POPCO engineers led county officials through a bewildering thicket of tanks, tubes, wires, valves, computers, gauges, and pipelines. It was the world's most modern gas processing plant, its managers said, and should make the county proud.

But the plant began operation December 28, 1983 with a near disaster. Deadly hydrogen sulfide, which the plant was designed to purge from the gas, got past the processing system and into distribution lines leading to several Santa Ynez Valley communities a few miles north and to President Reagan's ranch. The president was away, enjoying the holidays in Los Angeles. Most Santa Ynez Valley residents who were home were advised to leave, but most stayed put; the weather was mild and they just kept their gas jets closed. POPCO vented and purged the lines before anyone inhaled the toxic gas.

Several systems, including redundancy equipment and mishap detectors, had failed simultaneously. After a six-month shutdown for repair, redesign and adjustment, the plant operated without mishap. Work began in the mid-1990s to double its capacity to accommodate new offshore gas production.

Waterfront Resistance

While these impressive ventures took form, some big plans never materialized, including proposals for vast coastal supply bases, pipelines to Los Angeles refineries, tanker terminals, offshore platforms, and onshore processing plants. They were victims of financial shortfalls, environmental problems, land use restrictions, depressed oil prices, and other frustrations. While these endeavors were pending, however, they kept county planners busy trying to accommodate them within county, state, and federal policies and resources.

During the agonizing process, it was obvious that county policies were not adequate to cope with the industrial plans that did materialize. A comprehensive, enforceable plan for dealing with the *industry*—not just individual companies—was essential. But each company wanted its own processing plant, tanker terminal, storage tanks, and pipelines. The county planning staff, headed by Albert Reynolds, urged the Board of Supervisors in 1975 to begin work on a comprehensive plan. But not much happened. The supervisors would not approve funding for such planning, the many agencies involved could not agree on an approach, and the industry called any such attempt unwarranted public interference.

Between 1975 and 1985 the county managed to limit new or expanded onshore industrial plants to the six coastal canyons which already had oil and gas plants. Meanwhile, seven more oil production platforms appeared offshore. Santa Barbarans who felt threatened by this trend organized "Concerned About Oil" and placed a proposed ordinance (Measure A) on the November 1985 ballot to limit new onshore oil construction to one consolidated site on the county's south coast. The old plants would be phased out. Another oil complex could be built north of the Santa Ynez Range in the northern half of the county. Tankering would end by November 1, 1988, and one interim tanker terminal would serve until then.

Threatened with diminished discretionary authority, the Board of Supervisors placed an advisory Measure B on the same ballot to counter Measure A. In Measure B, the supervisors promised to limit new onshore construction to two south coast sites, encourage pipeline oil transport, and increase efforts to control air pollution. Measure A took a beating, and the advisory Measure B passed. But Concerned About Oil partisans felt they had caught the supervisors' attention and headed them in the right direction.

Other coastal communities also were paying attention. They watched Santa Barbara fight a rear-guard action against oil development, and they decided to put controls in place before they were invaded. Several cities and counties adopted policies that banned or limited onshore oil plans. In communities where political leadership was lacking, the people took the initiative. In November 1986, voters in nine California coastal cities and counties approved measures that prohibited onshore facilities, required a vote for such construction, or delayed construction pending a decision on a permanent ban. The San Diego city measure prohibited city employees from cooperating with the oil industry. All nine measures carried, most by two to one ratios. And that was just the beginning.

"Obviously, this is something that is spreading like wildfire," said Richard Charter, coordinator for several California local governments that opposed leasing off Central and Northern California. He said nothing had come of two years of negotiations with the oil industry and the Interior Department over where to permit offshore leasing. Now, he said, "California coastal jurisdictions are turning to the electoral process...to roll up the welcome mat and declare the coast off limits to drilling-related industrialization." Charter said he expected other communities would soon place similar measures on the ballot.[6]

He probably did not know how right he was. A Santa Cruz citizens group, Save Our Shores, began organizing other California coastal communities to resist oil industrialization. Inspired by its 29-year-old executive director, Dan Haifley, the group organized successful anti-oil campaigns almost from one end of the state's 1,100-mile coast to the other.

Between 1986 and 1990, a total of twenty-six California coastal cities and counties severely restricted building anything associated with offshore oil production. All such proposals would have to be put to a vote of the people. In seventeen communities, measures were enacted by popular votes which ranged from 53 to 85 percent approval. The average was 72 percent. Nine city councils and the San Diego County Board of Supervisors passed similar ordinances. The oil industry filed suit against several of the local actions, but the courts held that such decisions by the voters and/or their governing bodies were valid.

Why had Measure A failed in Santa Barbara County, the traditional center of offshore oil opposition, when so many other coastal communities succeeded? In Santa Barbara, the environmental community was spilt over the 1985 initiative. The Board of Supervisors had added a competing measure to the ballot. Some environmentalists felt that more could be gained by lobbying the Board of Supervisors than by "tying its hands." Beyond that, the oil industry was "organized" in Santa Barbara County; it had offices, contacts, and staffs. And it had rich, known oil deposits in nearby protected waters. It was willing to spend freely to defend this treasure from an adverse public vote. Because the Santa Barbara initiative was the only such measure on the ballot in 1985, industry effort was concentrated there. After that the industry's resources were spread over several brush fires at each election.

In any case, the Santa Barbara County Board of Supervisors moved ponderously but seriously toward a comprehensive plan for accommodating the oil industry while keeping resource damage to a minimum. After ten hearings in 1987, an integrated plan was approved. The industry received two locations for new onshore plants, and the eight existing complexes were frozen at their originally-authorized level of production. The county soon went beyond "freezing" these installations; in the early 1990s it moved to phase them out by rezoning the old sites to non-industrial uses. That made oil activity legal non-conforming which meant the plants could be maintained, but not renovated, enlarged, redesigned, or replaced. Two massive new plants were built; one almost out-of-sight in Las Flores Canyon, the other a conspicuous industrial complex at Gaviota along Highway 101, with an overpass linking it to a tank farm and tanker terminal on the oceanfront.

Unsatisfied even with this compromise, two companies in 1995 sought permits to slant drill under state waters from onshore, *outside* the two designated locations. Impelled into action by these proposals, environmentalists qualified a county initiative for the March 1996 ballot that would require a vote of the people to approve such permits. Mobil Oil waged a long campaign for its oil drilling plan on university land near the campus, but the university administration said, in effect, "not in my backyard."

Mobil finally abandoned its effort. Even before the election on the ballot issue, Molino Energy Company decided to relocate its natural gas slant-drilling operation in one of the designated oil industry locations.

Fishers Protest

While the county kept much coastal land free of industrialization, commercial fishers were trying to defend their ocean realm. Oil drilling vessels, each with cables reaching to several far-flung anchors, claimed much of the ocean that had been open for commercial fishing. Fishers said that the busy boats serving offshore oil operations often disregarded fishing vessels or trap markers. Drilling from a flotilla of vessels was followed by even more drilling from production platforms. This produced mounds of drilling muds and drill cuttings that smothered vast areas of oceanbottom habitat. Fishers also wondered about fish damage from drilling mud chemicals. (See Chapter Twelve). Ships and barges laying pipeline were also in the way, and the oceanbottom pipelines themselves were sometimes obstacles to commercial fishing.

Suddenly vast fishing areas were threatened by the oil industry. "We can lose an area of five miles radius around every [drilling] rig that anchors," said Alan Hur, a lobster and crab fisherman. After aggragating all overlapping space required by proposed oil projects in the Santa Barbara Channel, Gordon Cota, another commercial fisherman, told the state Coastal Commission, "I'm going to be 223 percent out of business." He did not deny a little facetiousness, but he wanted to emphasize that the oil industry required huge three-dimensional blocs of ocean at the expense of other enterprises that were there first.

In March, the Mediation Institute of Seattle and the University of California Cooperative Extension Sea Grant Advisor set up a meeting of the adversarial fishing and oil industries. One result was the Joint Oil/Fisheries Committee of South/Central California, or simply, the Joint Committee. It was to establish rules that brought both industries as close as possible to allowing two bodies to occupy the same space at the same time.

Seismic survey crews were to post the time and place of their operations, and fishers with set nets and oceanbottom traps were to do the same. Each would know where the other was at all times, and each could avoid conflict by abiding by the Joint Committee's "rules of the road." Routes were set for oil industry service boats to avoid churning through fishing gear. A Liaison Office, another creation of the April meeting, was charged with helping fishers file claims against oil company boats or other operations that damaged their fishing gear or reduced their catch.

The efforts of the Liaison Office did not eliminate conflict between the oil and fishing industries, but it reduced it substantially as a clearinghouse where the two could communicate. They developed rules intended to avoid conflict and to compensate for damages.

But the conflict with seismic surveys continued. It began in the 1950s with marine dynamiting and quickly became a highly sophisticated technology, but fishers still found it bothersome. In the 1960s, the oil industry stopped dynamiting the ocean waters as a way to delineate the geologic structure beneath the sea floor. This was now achieved by shooting acoustic pulses from "air guns" and recording the sonic reverberations in electronic equipment aboard research vessels. This gave the companies a "picture" of the subsea formations. The industry maintained this system did not bother fish. Commercial fishers were not convinced. If the sonic bursts were not lethal, the noise and the jolt dispersed the fish, they said, and no one knew what the air blasts did to fish larvae.

The oil industry meanwhile insisted that "there is no danger to fish resulting from modern-day seismic survey vessel operations," but acknowledged that whether the blasts dispersed fish was not resolved.[7] In March 1984 a scientific workshop concluded that more study was needed on fish dispersal. Before the end of the year ARCO contributed the use of its new 247-foot *ARCO Resolution*, a floating laboratory, for a week of tests to study the reaction of fish to sonic air blasts. It was one of several research projects, including one funded by the Minerals Management Service, that studied the matter but did not resolve the dispute nor end confrontations at sea.

Cables, sometimes two miles long with hundreds or thousands of hydrophones were towed by research vessels. When the ships made a turn, the cable followed in a long sweeping arc, sometimes tearing loose or dragging buoys that marked crab and lobster traps or set nets. Fishing boats were required to give way to approaching seismic survey vessels with their long cables in tow. The research vessels could not change speed nor direction on short notice. But fishing boats with large nets or other unwieldy gear in the water were hardly prepared to get out of the way in a hurry.

In 1995, Exxon received federal permission to resume seismic testing in the Santa Barbara Channel via sonic bursts. The company had to steer clear of migrant whales, scale down the force of the blasts, and cease testing by the end of the year. Given marine mammals' strong reliance on sonic communication and the possible vulnerability of their "hearing" organs, the conditions Exxon accepted did not evoke total public confidence.

Everyone agreed the conditions were temporary and based on incomplete information. In 1996 the Minerals Management Service organized a team to craft procedures for reviewing seismic surveys permit applications. The High Energy Seismic Survey group, including people from public agencies, industry, research, academe, and environmental organizations, was still at it late in 1997. By that time the technology was incredibly sophisticated and the marine gear immense. It was difficult to see how it could be applied in an area as small, and as biological rich, as the Santa Barbara Channel without unusual environmental repercussions.

Senators Edmund S. Muskie and Alan Cranston (bending) and a Coast Guard officer examine the oil-soaked beach in the Santa Barbara harbor in 1969. (*Santa Barbara News-Press*)

Chapter Twelve

Contaminating the Water and Air

The offshore oil industry has dumped a lot of waste into the ocean, much of it quite legally. The rest was dumped illegally, in innocence, accidentally, or in some combination of culpability. All of it degraded the water, and the debate over how much should be allowed probably will persist longer than the industry.

Discharges come in the form of used drilling fluids, drill cuttings, water that comes up with the oil, wastewater and other gunk from operations on drilling rigs and platforms, oil spilled accidentally (or dumped surreptitiously), and chemicals separated from oil and gas, some of which are devastating.

Public agencies set standards for waste disposal, but they usually played catch-up. The industry wanted to go into action before regulators knew how much damage was done by each kind of waste. Initial standards often were quite arbitrary, then revised—usually belatedly—as government learned more about what really happened underwater. Public agencies then crafted enforcement techniques, many of which the ubiquitous "concerned citizen" viewed with little confidence. Government and industry also sought ways to contain and clean up oil spills, and to decide who was responsible for cleaning them up.

Drilling and Dumping

Serious pollution problems begin with drilling. When you drill or dig a hole, you have to find a place for what you remove. Drillers have more to dispose of than what they gouge out of the ocean bottom. They circulate drilling fluids through the well. These can be reused several times, but eventually they become contaminated and must be disposed of and replaced.

These fluids—often called muds—are pumped into a well through the drill pipe, past the drill bit, and then back up outside the drill pipe along the outer wall of the well. Among other functions, drilling muds lubricate and cool the drill bit, control pressure in the well, and carry drill cuttings to the surface.

"Most drilling fluids are muds which range in composition from slightly dirty water to water containing a very high concentration of clays, weight materials, soluble salts, conditioning agents, and emulsified oils."[1] The principal bulk constituents are water, barite (barium sulfate), clay minerals, chrome lignosulfonate, lignite, and sodium hydroxide.

Together with the drill cuttings, the discharged muds bury a part of the ocean bottom and "smother" communities of benthic (bottom dwelling) creatures. How long this dumping continues and how far this blanket of waste extends depends on how long each drilling operation continues (weeks, usually) and how many wells are drilled in a limited area. Production platforms have slots for as many as ninety wells.

In the early 1980s, it was estimated that nearly 200,000 metric tons (dry weight) of drilling fluids were being discharged annually in federal waters off California.[2] The Environmental Protection Agency in 1985 estimated that as much as 1.5 million barrels of drilling muds, fluids and other waste products could be expected to be discharged into federal waters off the state's coast from future oil development.[3]

That would create numerous large mounds to be recolonized to replace organisms destroyed underneath. But the "smothering" was not the only—or even major—concern. Even if drilling muds were only slightly toxic initially—a widely-disputed claim—they were not that benign when pumped into oil wells. Each company added chemicals to make the mud more effective for specific drilling operations. What did these additives do to marine organisms? A National Research Council panel stated in 1983, "there is limited information" on these additives. "Several common drilling-fluid additives, including biocides and diesel fuel (No. 2 fuel oil), are much more toxic to marine organisms than the bulk constituents."[4] There was little reliable research on the effect of drilling muds—with all their constituents—that were dumped into the many and varied ocean environments. Compounding the problem, companies often maintained that what they added to the muds was nobody else's business; a trade secret.

With not much to go on, the Environmental Protection Agency in 1967 set toxicity standards for Platform Hogan, the first platform in federal waters off California. As the federal government learned more about waste discharges, the EPA in 1978 tightened dumping standards for Platform Hogan and Platform Houchin approximately four miles off Carpinteria and Summerland.

This trend toward stricter standards was immediately reversed by the Reagan administration. "We have a reputation for being a tough region," said Sonia Crow, new head of EPA's Region IX on the Pacific Coast. "But those days are over," she told the oil industry. "The door is now open. We are interested in your concerns and problems. We do not regard you any longer as an adversary against whom we must protect the environment."[5] In her address to the Western Oil and Gas Association in Los Angeles, she said EPA would no longer review discharge permit applications on a case-by-case basis, but would use a single set of standards for all permit applications.

Public concern deepened over what was happening out of sight in the ocean environments as offshore drilling rapidly increased in the late 1970s and early 1980s. As several five-year permits were about to expire, EPA extended their terms to gain time for more research and public hearings.

EPA established a single discharge standard for the Santa Barbara Channel and Santa Maria Basin. Critics said this made no sense for a nearshore region that included several kinds of ocean bottoms, current patterns, water temperatures, marine organisms, and varied habitats. They did not want the general permit extended while new rules were studied, and above all did not want another general permit under the new rules. At several meetings in 1983, critics challenged the oil industry and the Interior Department which insisted that the discharges were virtually harmless.

The industry provided seven experts in support of extending the blanket permit. Their testimony was summarized by Robert Ayers Jr. of Exxon who said that drilling muds have "low toxicity...approximately 1,000 to 10,000 [times] less toxic than common household detergents." Ayres, a research associate for Exxon, participated in many discharge toxicity studies during the 1970s and 1980s.

Local agencies and non-governmental scientists protested that the waste did far more damage than the industry or Interior Department were ready to acknowledge. By this time Dr. John Mohr, a retired professor of marine biology at the University of Southern California, was building a reputation as an offshore environmental gadfly. He followed waste discharge research closely, and frequently publicly protested against what he considered faulty methods, incompetent researchers, and unwarranted conclusions. He said that oil companies and the federal government used "low quality" scientific information on the effects of drilling muds, when better information was available. He was rapped out of order when he jolted a hearing officer from his torpor by implying a "connivance" between the petroleum industry and Interior Department in using data that favored the oil companies.

The Environmental Protection Agency extended the general permit for six months, and that became a year and a half. Then it returned to Santa Barbara for more hearings on whether the new permit should be general, or separate and specific to each drilling operation. This time the EPA heard more scientific testimony opposing general permits. Among those objecting were Alice Alldredge, associate professor of marine biology; Daniel Morse, professor of genetics and biochemistry, and Richard Zimmer-Faust, marine biologist, all of the University of California at Santa Barbara. Because of the varied environments, habitats, and marine populations, the UCSB scientists recommended setting standards case-by-case.

The industry was represented by John Hallett of Shell California who indicated he spoke for fourteen companies of the California Offshore Operators Ad Hoc Committee. The companies, he said, had found "some serious substantive problems with the permits," which would ban discharge of diesel oil [a drilling mud additive] and limit the discharge of drilling mud containing mercury and cadmium. He argued the oil industry opposed limits on dumping drill muds "in the absence of a scientific showing" that they have a significant adverse impact on marine life.

This would appear to shift the burden from the industry to show that its discharges were harmless or acceptable, to the public to demonstrate that they were harmful. Ordinarily, one must prove a benefit or at least no unacceptable harm, from an activity requiring a public permit. In this case, the Environmental Protection

Agency suggested stricter dumping standards, but stuck to the concept of a blanket permit.

WHAT DID THE STUDIES SHOW?

While the damage of oil industry waste in ocean waters was poorly understood, such waste was accumulating rapidly in the Gulf of Mexico and off the California coast. Consequently, in the late 1970s and in the 1980s the government, industry and scientists became serious about testing the waters.

In the early 1980s the National Research Council assembled a panel to assess "fates and effects of drilling fluids and cuttings in the marine environment." Even before the panel reported, it was attacked as stacked in favor of industry. The panel prepared "working papers" which critics said verified their fears. "To be blunt, I believe that a miscarriage of science is in the making," said Elliot A. Norse, director of science and policy for the Center for Environmental Education. In a letter to Frank Press, president of the National Academy of Sciences (parent to the National Research Council), Norse wrote, "I believe that the criticism of the panel and its working papers is justified." [6]

As far as California was concerned, the Governor's Office of Planning and Research maintained that it did not see much relevance in the working papers because they had almost nothing to do with Pacific waters. There were no Pacific coast marine scientists on the panel which held all its meetings on the East Coast. The state argued that California experts were disturbed because the panel used unreliable previous studies. Finally, it said, "the panel is composed of several members who either work for the oil industry or have worked for the industry in the past."[7]

In a letter to John Mohr, who found the working papers unreliable and the panel composition unbalanced, Frank Press said that "some of your concerns are shared by others....[S]everal individuals have written me concerning the perceived lack of balance on this panel....I intend to dispel this perception by adding additional well-qualified individuals to the panel."[8]

Press said that "On the matter of quality of the panel's work, it is premature to judge" on the basis of the working papers. But critics felt the final report was not much of an improvement.

California marine scientists had argued that a single standard for waste discharges could not be justified for the waters off Ventura and Santa Barbara counties because of varied environmental conditions. Certainly, then, they would not accept extrapolations from Atlantic and Gulf studies to Pacific waters. The National Research Council panel report itself seemed to disclose ambiguity:

> It is desirable to tailor drilling-fluid regulations to take account or advantage of environmental conditions or to protect sensitive or valuable habitats, but there is no evidence that justifies different regulatory policies concerning the use of drilling-fluid additives in different geographic regions.

But the next sentence says,

> Discharges of drilling fluids and cuttings into OCS waters take place in a wide range of marine environments, which vary greatly in water depth, ice cover, tidal and nontidal currents, waves, geological history, land runoff, and biotic characteristics. Thus, the physical fates of discharged drilling fluids and cuttings vary greatly.[9]

During hearings in Santa Barbara in 1983, government officials and industry executives spoke as though all the necessary information was at hand to confidently set discharge standards, and at a level lower than recommended by scientists not associated with government or the oil industry. Reviewing the results of research conducted to that date, it appears that there was still much to be learned, that the Interior Department and the oil industry heavily influenced much research, and that federal agencies tended to rely on research conclusions that most nearly accommodated oil production.

Research continued throughout the 1980s, and it disclosed how much was not known about the damage of waste dumping on marine life. Studies now included Pacific waters, but the results were often inconclusive. Some research was criticized as unreliable, some competent studies disclosed that discharges were more detrimental than once believed, and still others concluded that more study was needed.

Such research was difficult because of the number of toxins and other materials discharged, the variety of habitats, the vast number of organisms in the ocean, conflicting results from different test methods, conflicting interpretations from the same data, conclusions reached from incomplete data, and the pitfalls of extrapolating short-term findings to long-term consequences, laboratory results to field conditions, and single well-drilling to multiple drilling operations.

Because its offshore oil operations were within three miles of the coast, California was more concerned about what was dumped into the ocean. The state required that drill muds and cuttings be disposed of onshore. That just changed the location of dumping, but waste onshore usually stayed put in safe places where you could keep an eye on it.

In the mid-1990s the Environmental Protection Agency again upgraded its discharge standards, but still permitted dumping drilling muds and drill cuttings that were within limited concentrations of particulates, cadmium and mercury. It continued to prohibit discharging waste which contained oil.

Discharge standards were one thing; enforcement was something else. The Environmental Protection Agency relied on self-monitoring by the oil companies which were reluctant to report any mishap they could cover up. "The requirement that permittees conduct self-monitoring of discharges is authorized by Section 308 of the Clean Water Act and is a standard requirement of all National Pollutant Discharge Elimination permits issued by EPA," the agency responded to critics in 1986. The question, however, was not whether self-monitoring was authorized. The issue was whether such a system inspired public confidence. EPA maintained the system was "effective and efficient." But at least two major discharge violations by Chevron and Texaco in the Santa Barbara Channel were uncovered in the 1980s only after whistle-blowers alerted the EPA.

Public confidence suffered further deterioration when the Environmental Protection Agency left inspection and enforcement to the Minerals Management Service, compounding MMS's conflict of interest in promoting as well as policing offshore oil activities. The EPA had neither the funds, personnel nor helicopters to inspect and police offshore operations. The Minerals Management Service had those resources because it was charged with regulating many other activities on the platforms. Santa Barbara County urged third-party inspections.

EPA and MMS did, however, strengthen enforcement off California. Because there were only a couple dozen platforms off the Pacific Coast, and because they were near shore, each was inspected far more frequently than the hundreds of distant platforms in the Gulf of Mexico. Moreover, California—not the Gulf Coast states—was the squeaky wheel. MMS often inspected California's platforms on short notice. There had to be *some* notice because platform crews had to prepare for helicopter landings.

COPING WITH OIL SPILLS

The best thing to do about oil spills is prevent them. After 1969 the oil industry vastly increased the reliability of its technology and the competence of its crews. It helped, but no system is foolproof, and much of the pre-1969 industrial plant—including pipelines, tankers, drilling equipment—was still in place and operating. The safety record improved, but not enough to prevent many small accidents and some devastating spills. Public agencies drafted contingency plans to deal with spills quickly and effectively. The plans were detailed, specific, and comprehensive. But they almost never worked as well on water as they did on paper. The plans were quickly outdated; phone numbers, addresses, organization, responsibility and personnel changed on the carefully-crafted contingency chart. Some people did not do what they were supposed to do, some did not *know* what they were supposed to do. Emergency equipment too often was not kept in working order.

In the case of the *Exxon Valdez* grounding, some major equipment was in drydock, not on active standby. One oil spill cleanup vessel left Santa Barbara County waters for Prince William Sound without authorization, leaving a protective gap off California and providing no useful service in Alaska.

After the tanker *Puerto Rican* exploded ten miles off San Francisco Bay in 1984, currents carried the oil slick in a direction unanticipated by the contingency plan. Three years later two freighters continued on a collision course ten miles off Point Conception because radio operators on an oil platform could not rouse either crew. Neither vessel was a tanker, but the *PacBaroness* sank and for months polluted the waters with diesel fuel and chemical sloughing from its cargo of copper.

Two years after the 1969 blowouts, fourteen oil companies operating in the Santa Barbara Channel organized a co-operative, Clean Seas Inc., and assembled the best cleanup equipment available. That included mostly floating barriers, or booms, designed to contain oil slicks. The state and the county asked the co-op to demonstrate its efficiency in drills—preferably unannounced—but the co-op resisted, preferring to make its claims in colorful brochures. Surprise drills, Clean Seas said, were designed only to embarrass the industry cleanup cooperative.

Some tests, however, were witnessed by public officials. In 1972 soybean oil was deliberately spilled ten miles off Point Conception and a 1,000-foot plastic curtain was deployed which contained most of the 25,000 gallons. The Coast Guard wanted to evaluate the effectiveness of the equipment in five-foot waves and 20-knot winds. But the weather did not cooperate; it was a beautiful day with swells of one to two feet, and no wind. In earlier tests—without soybean oil—the barrier was heavily damaged by ten-foot seas and winds up to thirty-five knots.

The industry conducted other tests on its own terms, with and without public audiences. In one case it contained a popcorn spill impressively, according to a Lean Seas spokesperson. The floating barrier rounded up the popcorn in four to six foot seas, the executive director said, and from that he extrapolated that it would contain oil in six to eight foot seas. Clean Seas literature reported that some barriers were "designed to survive in waves up to 20 feet high," but survival was not the issue; containment was, and obviously oil could not be contained in twenty-foot seas.

In a region where natural oil seeps could have been a convenient laboratory for testing oil containment gear, why were all the tests in clean water, or with soybean oil and popcorn? Clean Seas officials explained that seep oil gummed up their equipment and was difficult to remove.

Exuberant claims for cleanup gear diminished public confidence in what was supposed to protect natural resources and coastal urbanization from oil spills. The Coast Guard and the California Coastal Commission, which also called surprise tests, concluded that spilled oil could be contained in calm seas, with a current of no more than one and a half knots, and little or no wind. In strong winds, currents, or waves, oil splashed over or flowed under floating barriers.

The oil industry, relying on its own engineering genius and prodded by regulatory agencies, developed the best possible technology for containing spills. It improved the booms and shortened the time required to deploy them. In 1981, it launched a 136-foot workhorse of a vessel, *Mr. Clean*. It was big, powerful, loaded with oil containment and cleanup equipment, and staffed by a trained and ready crew. Moored off Santa Barbara, it was the first of three similar ships to be stationed in the area.

The industry, if not the public, began to feel comfortable about the safety and tidiness of the oil industry. Early in 1989 the Santa Barbara County Energy Division recommended to the Planning Commission a massive set of requirements to be imposed on oil companies which wanted to construct a marine tanker terminal at Gaviota. The Energy Division maintained that the industry was grossly unprepared for potential disasters. Industry spokespersons, insulted and outraged, insisted that they were prepared to cope with any tanker emergency. The vote was delayed for a month for further consideration.

When the oil men returned to the Planning Commission for the second hearing, their mood had changed. In the interim, the *Exxon Valdez*' grounding had demonstrated the haplessness of contingency plans and containment technology in a major tanker spill. The industry was prepared for emergencies, the oilmen said, but of course they did not mean *that* kind of emergency. They continued to insist, however, that the county had underestimated the industry's readiness to cope with lesser spills.

As the 1969 blowouts had stimulated action for greater oil spill protection off California, the *Exxon Valdez* disaster now demanded increased protection on all the nation's coasts. Three months after the *Exxon* grounding, the American Petroleum Institute announced "a new nationwide industry program for prevention, containment and cleanup of major oil spills." It would create a Petroleum Industry Response Organization (PIRO) with five regional response centers "prepared to act quickly in any future major spills." It would also recommend improvements in tanker operations and would fund research and development for oil spill control and mitigation.[10]

A year later PIRO was replaced by the Marine Spill Response Corporation (MSRC) with a similar program. PIRO had planned to have one of the five bases in Long Beach; MSRC decided instead on Port Hueneme in the Santa Barbara Channel. Under each regional response center would be one to six smaller bases with oil containment equipment. In its initial announcement, the Marine Spill Response Corporation was frank about how much protection it could offer:

> In adverse weather, it is often impossible or too dangerous to allow equipment and personnel to leave the shoreline even when the equipment is locally staged and ready to begin the response....Other factors that affect the ability to clean up a spill include the type of accident, its location, the rate of discharge, the time of day, the season, currents and tide, oil characteristics, regulatory approvals, inherent equipment limitations and many other physical and operational factors. Thus, MSRC cannot promise to remove oil in all circumstances, or to prevent oil from reaching the beach.[11]

The *Exxon Valdez* grounding obviously was a sobering experience!

Removing spilled oil from the water was not the only way of getting it off the surface; you could also sink it, or at least try. That would not get it out of the water; just out of sight and prevent it from reaching shore. For this, the petro-chemical industry produced dispersents which are solvents or other agents that break up surface tension that supports oil on the water. Remove this support and the oil sinks.

This can protect sensitive shorelines and coastal habitats. But it does not always work, and the dispersant themselves are toxic. If that degree of toxicity is preferable to the harm that might be caused without it, the damage is limited. That is a tough call when there is little time to consider alternatives.

In the turmoil that usually accompanies major oil spills, the use of dispersant often is botched. Sometimes it is applied in the wrong place, especially when spread from aircraft. Sometimes it is the wrong type of dispersant, or maybe the oil is not dispersible. Crews using the chemicals must know how to apply them and be able to select the right dispersant for the job. Often it was impossible to evaluate how effective a dispersant was because the results were not monitored accurately.

Dispersants and the way they are applied have improved, but twenty years after they were used ineffectively in the Santa Barbara Channel blowouts, a National Research Council study concluded that, "[f]ew published studies exist on the fundamental science concerning how dispersants act on oil in water." It did report, however, that "[a] wide range of sublethal effects of dispersed oil has been observed in the laboratory."[12]

There's Something In the Air

While California's fear of oil spills was understood worldwide, the state was equally distressed about what the oil industry was doing to its atmosphere. Coastal California depended on blue skies and fresh air for its economic and public health. Much of the state's coastline, from San Francisco to the Mexican border, suffered from photochemical smog, and petroleum was responsible for nearly all of it.

In summer and autumn when auto traffic is heavy and the winds balmy, coastal plains are poorly ventilated under a blanket of smog. Daytime onshore breezes carry it against a mountain barrier, and nighttime offshore winds move it back over the same area.

Before anybody knew what this airborne irritant was, Southern Californians coined "smog" by combining the words smoke and fog. It was neither, but the name became so common (and convenient for headlines) that when Dr. A. J. Haagen-Smit discovered its true nature in 1950 all scientists could do was add "photochemical" as a modifier.

Strictly speaking, the oil industry does not produce smog. It produces most of the oxides of nitrogen and reactive organic gases that form ozone when exposed to sunlight. Ozone, the main component of photochemical smog, is a more nearly accurate name than smog for the pollutant.

As usual with new and strange industrial-chemical afflictions, the scourge was well upon its victims before much could be done to moderate its consequences. After scientists discovered what smog is and whence it came, it began to dawn on Californians that the oil industry on which the nation had built much—if not most—of its economy, was assaulting its beneficiaries.[13] Most of the smog was produced by automobiles, and some political leaders were ready to take direct action. In 1969 and again in 1970, the California Senate passed bills that would ban automobiles powered by internal combustion engines. This visionary leadership did not extend to the California Assembly where the bills died.

Lesser steps, however, were enacted. The 1970 federal Clean Air Act and similar California legislation designed to limit auto exhausts and other sources of petroleum pollution were enacted over the protest of the auto and oil industries.

In California, counties or regional air quality districts were responsible for devising policies to meet federal and state standards. But they lacked authority in some important areas, such as auto emission control which was a federal responsibility. Congress did allow California, however, to add stricter controls than required by federal law. Some Southern California counties started from a deficit. They were not required to *maintain* air quality; they were already in noncompliance and had to work out controls that would *achieve* required standards. Eager to uphold its reputation as environmentally undefiled, Santa Barbara County took air quality control seriously. But it was caught between the Environmental Protection Agency which demanded compliance, and the federal Minerals Management Service which undermined county efforts. MMS air quality controls on oil operations in the Outer Continental Shelf—beyond local control—were so lax that air pollution from offshore overwhelmed county efforts to meet EPA standards.

That's your tough luck, the state and federal governments told Santa Barbara County. William Lockett, chief planner for the California Air Resources Board, was asked in 1978 how the county was expected to control pollution from the OCS where it had no authority. There are overriding national interests that must be accommodated, he said.

"Do you mean that we must reduce our automobile use, for instance, to make up for the additional pollution we get from offshore tanker loading?" asked Leland Steward, county transportation director.

Lockett said that standards must be met without regard to the source of the pollution. "There could be pollution from activities outside your jurisdiction from which you might gain benefits," he said.

John English, county air pollution control director, joined the fray: "How are we to achieve standards when activities in the federal waters off our shores are getting exemptions to emit four times the amount of pollutants from each source than those within the county?" he asked. "How can we develop strategies to offset the impact imposed on us from areas we can't control?"[14] The question went unanswered for twelve years.

But it kept being asked. At a news conference in Los Angeles in 1982, I asked it of Anne Gorsuch, President Reagan's EPA administrator. The source of the pollution "is not our concern," Gorsuch replied. "Our concern is whether or not the area is in attainment." I persisted:

> The [county] Air Pollution Control District says that the county can meet federal attainment standards, except that the pollution from OCS federally permitted activities will undermine it. Have you been in contact and consultation with Interior Department agencies over this apparent conflict?

"There is really not a conflict," she insisted. "It's not a matter of which activity will cause a non-attainment status. That's really up to the local jurisdiction to determine."

She was told that these activities are virtually beyond any local control. She replied:

> It's a typical argument....'We're not doing it here—we're receiving everybody else's problem.' I don't know the particulars on that [Santa Barbara situation]. Let me just say that regardless of the kind of activity, the singular consideration of EPA is, will there be attainment. Forget whose industrial activity it is. If there is not attainment the law requires [EPA] to pursue a course of action that results in sanctions [penalties].

I tried once more: "But if there is not attainment in this case, it is the result of a federally-permitted activity over which neither the state nor the local government has any control." Gorsuch replied: "Wherever there is an industrial plant—these plants operate under federal government permit...it's not our concern. Our concern is whether or not the area is in attainment."[15]

At that time I was convinced that Gorsuch did not understand the question, or the problem. Later events, however, indicated that this was deliberate federal policy. The oil industry was granted loosely regulated use of natural resources, including clean air, leaving local governments to cope with the consequences, an impossible task under the circumstances.

At the December 1982 Western Oil and Gas Association conference, Gorsuch told California counties to meet standards and deadlines set by the Clean Air Act or face the loss of federal funding for projects such as highways and sewage treatment plants. Critics said it was part of a plot to force Congress to weaken the Clean Air Act. She said it was simply her duty to enforce the act. Instead of weakening the act, Congress objected so vocally to her management tactics that she was forced to resign three months later.

Decade of Frustration

Throughout the 1980s the counties, consultants, state and federal agencies and the oil industry wrestled with federal offshore air quality control. Coastal communities argued air pollution from oil operations on the Outer Continental Shelf was an increasing threat. The Interior Department and the industry maintained it amounted to almost nothing. EPA, caught in the middle, agreed it was a serious air quality threat, but the agency lacked authority to set emission standards in federal waters. Neither side could agree on how to determine the onshore impact of OCS pollution.

Despite its lack of authority on the OCS, however, the county was not helpless in controlling air pollution from offshore.

"The principle handle that the county used was the Coastal Commission consistency review," said William Master, then assistant director of the Air Pollution Control District. The federal Coastal Zone Management Act authorized the California Coastal Commission to object to federally-permitted activities offshore if they conflicted with the state coastal plan. "It became a major federal issue because the Department of Commerce had authority to review and overrule the Coastal Commission's determination of consistency [with the state coastal plan]," Master said. "A number of those projects did go on appeal to the Department of Commerce along with extensive lobbying by both sides to try to influence Commerce's decision. So it had very high visibility including, I believe, all the way to the White House."[16]

Another county tactic was described by Kathy Milway, intergovernmental affairs manager for Santa Barbara County's Air Pollution Control District:

> The county was taking the strong stand that we're not going to authorize anything onshore that's tied to the OCS unless you [Interior Department] clean up the OCS. Those kind of negotiated permits were driving companies crazy because they did not know where they stood in the permitting process. There was nothing written down that said these are your requirements. They did not know if they were responsible to the Minerals Management Service or to the county. In order to get their permits they were certainly interested in negotiating with the county....MMS started hearing the problem and undertook something called negotiated rulemaking.[17]

She said MMS sought to change its regulations in a way that would preserve its exemption from the Clean Air Act.

Negotiated rulemaking was suggested by Interior Secretary Hodel in 1985. That was an attempt to coordinate air control rules with a number of agencies involved. He eventually agreed to include Santa Barbara and Ventura county officials, and later even Santa Barbara environmentalists. John English, head of Santa Barbara County's Air Pollution Control District, said this procedure would only delay effective action which he said was needed immediately.

"What's the bottom line?" he asked of Hodel's schedule. "A year? 18 months? That's too late for us."

A long line of companies had applied for offshore production permits. English repeatedly warned that the county was running out of offsets which had to be provided onshore to compensate for pollution created offshore. As it turned out, two and a half years elapsed before the negotiations collapsed.

Santa Barbara County suggested certain levels of control for the Interior Department to impose on Exxon's big Santa Ynez offshore development. The federal department rejected most of the county's recommendations. Meanwhile, Exxon charged that the county's attempt to impose strict controls onshore was in effect an illegal attempt to extend its authority into federal waters beyond its jurisdiction.

Counties to the south were also vulnerable. Ventura County, which faces on the Santa Barbara Channel, was not only a target for pollution from offshore activity, but its deepwater port at Port Hueneme got heavy marine traffic from the oil industry. Ventura County failed to meet EPA's air quality standards, largely because of oil industry pollution. In mid-1986, Ventura County's ozone pollution was third-worst in the state and sixth-worst in the nation. In Los Angeles County much of the bad air came from oil tankers and refineries. Many of its leaders were not eager for more of this kind of business unless it was cleaned up. Marvin Braude, then a Los Angeles city councilman, was incensed in 1986 when the Reagan administration again gave the auto industry additional time to meet federal auto mileage standards.

"It is hypocritical for the Interior Department to push for offshore oil drilling (with its own significant air-pollution impacts) while the Highway Traffic Safety Administration ignores conservation measures," Braude said.[18] Bigger cars burned more gasoline per mile, but the big money was in selling big cars.

When Governor Deukmejian's Environmental Affairs Agency came to Santa Barbara in March 1986 to hear comments on federal plans for more offshore oil leasing, thirty-three officials and private citizens from Santa Cruz in the north to Los Angeles in the south testified. They said they were concerned that they would be—if they already were not—shrouded in contaminated air from a source they could not control.

It is not that they did not try. Under pressure from the state and EPA to do more, and the auto and oil industry to do less, county and regional air quality control agencies struggled throughout the 1980s to find ways to reduce pollution. Some oil industry apologists complained that the industry was taking too much blame for bad air. Autos, they said, also dirtied the air. But they did that with the oil industry's primary product. When you add it all up, petroleum was at the bottom of nearly all ozone production. And ozone is the chief constituent of what we call photochemical smog.

In 1990 in Santa Barbara County, the oil industry accounted for more than eighty percent of all oxides of nitrogen and other reactive gases that are precursors of ozone. This included pollution from production, transportation, marketing, and other oil commerce, and fuel combustion in stationary and mobile sources.[19]

This illustrated the county's dilemma. Two major polluters were essentially beyond county control. The Interior Department regulated pollution sources on the Outer Continental Shelf, and the state was in charge of policing auto exhausts. The state, however, was usually more concerned and cooperative, sometimes demanding more in auto smog controls than was welcomed at the local level.

The Air Pollution Control District tried to keep up with increasingly stricter standards accompanied by greater doses of pollution. It issued its first plan in 1979, and updated it with a 1982 plan that predicted the ozone level would be down to required levels by 1984. It did not happen, and the Environmental Protection Agency required a new plan by 1989. This one had fifty new regulations for the southern half of the county which suffered from ozone excesses. Then the state and the EPA demanded similar controls for the entire county.

Conservative county supervisors and much of the business community protested each new control. But the Air Pollution Control District explained:

> Failure to meet the requirements of the Federal Clean Air Act can lead to federal intervention in our local air pollution control program. Loss of federal highway funds, very stringent standards for new sources of air pollution, and increased control on existing businesses are some of the sanctions mandated by the Act.[20]

By now Bill Ryerson, who succeeded John English as APCD director in 1987, had built an "empire" of more than one hundred employees with an annual budget approaching $10 million. Critics found this a tempting target, and took aim. Why, they asked, did Santa Barbara County need such a massive and costly agency? They sought to compare it with similar agencies in other counties. But there *were* no similar agencies. Santa Barbara County was unique in the concentration of offshore oil activities that had to be accommodated and regulated. Most of the APCD's budget was funded by fees imposed on the oil industry. That did not assuage county supervisors who were under pressure from the industries who paid the fees.

A Congressional Coup

After several congressional and administrative efforts failed to tighten up offshore air quality rules, attempts were made to transfer this regulatory authority from the Interior Department's Minerals Management Service to the Environmental Protection Agency. Senator Pete Wilson in 1987 introduced legislation that would do the deed, but opposition from Interior and the oil industry overwhelmed any congressional support in evidence at that time. But at the end of the decade the federal Clean Air Act would be up for reauthorization.

"In discussing the reauthorization, the [transfer-of-authority] issue was raised by this [Santa Barbara] county as an important opportunity to preempt the Outer Continental Shelf Lands Act provision that excluded the Outer Continental Shelf from the Clean Air Act," said William Master, then assistant director of the Air Pollution Control District.

The opening was provided by Congressman Henry Waxman of Los Angeles who introduced a bill to transfer offshore air pollution regulatory power to the Environmental Protection Agency. County Supervisors Bill Wallace and Tom Rogers, accompanied by Master and Milway of the county staff, made cross-continental flights to arouse congressional interest. Most of the lobbying was up to Master and Milway.

> We wore out a lot of shoes....We did not know what we were in for. People told us later, 'We were laughing; [we said] you guys never should have tried it.' While we encountered reluctance initially by many staffers to even meet with us, once we were able to get them to sit down, five-minute meetings turned into 15 minutes, 30-minute meetings into one-hour meetings, and then requests to come back that evening.

Master and Milway made it a bipartisan, national issue instead of a partisan Santa Barbara matter. They said offshore oil was crowding out small business by using up all the pollution allowances. "Our pitch was the need for equity [for all businesses]...and that was bipartisan," Master said.

The Santa Barbarans had three goals: Requiring the same air pollution controls on the Outer Continental Shelf as were imposed onshore and in state waters; transferring regulatory authority from Minerals Management Service to the Environmental Protection Agency; and then delegating that authority to the adjacent local air quality agency.

Crucial as the issue was to Santa Barbara and a few other counties, it was a minor part of the Clean Air Act reauthorization debate. "These provisions were contained in what was called the "miscellaneous provisions" of the act," Master said. "That gives you an idea of the significance [attached to it]." Yet the sparring over the issue among legislators, staff members, Interior officials, oil lobbyists and the Santa Barbarans was intensive, and in the end Santa Barbara nearly lost it all. Master said:

> At one point in conference, because there was a tremendous amount of political pressure, and so many unresolved big issues, there was discussion of excluding the miscellaneous provisions from the act. The House had passed the provisions overwhelmingly, but Congressman John Dingell of Michigan who headed the House delegation in conference, opposed the House action. And the Senate had passed a weak version that did not call for transfer of authority.

Master maintained it was largely Waxman's intensity and skill that brought the vital amendments from the brink of abandonment to adoption by the conference committee and passage by both houses.

The victory was historic, but its impact was delayed. Major legislation is usually a policy statement. Nothing happens until the implementing rules are adopted. In this case it took three years for the Environmental Protection Agency to draft and publish final regulations.

Despite the accomplishment, the county's Air Pollution Control District victory was not celebrated by a conservative majority which was seated on the Santa Barbara County Board of Supervisors after the 1992 elections.[21] It attributed the economic recession (which was statewide) in Santa Barbara County to local overregulation of business. APCD's director, Jim Ryerson, was the primary target. Critics said his efforts to meet state and federal air quality standards were oppressive, too costly, and were driving businesses out of the county. But APCD officials noted that "study after study has shown that the costs for us to live with the pollution are much greater than the cost to control pollution at its source."[22]

But the new majority on the Board of Supervisors took a dim view of clean air controls, and as part of government reorganization plan lumped the Air Pollution Control District with Environmental Health Services and the Agricultural Commissioner's Office. Although each had highly technical and unrelated responsibilities, all would be under one super-director. It would not be Ryerson. Many accused the board majority of upsetting vital county government functions to get rid of one department head.

The deed was done in June 1993, and six months later the conservative Coalition of Labor, Agriculture and Business (COLAB) offered a new proposal for improving air quality. It suggested removing the monitoring station that recorded the worse air quality of the twenty-five monitors in the county. That would reduce the number of days in which part of the county exceeded the federal and state ozone standards. The station was inland between the north county and south coast in a sparsely populated area. COLAB blamed much of the hydrocarbon emissions on vegetation in Los Padres National Forest. The Air Pollution Control District maintained it drifted in from sources some distance from the monitor.

"As long as the Paradise Road [monitoring] station is there, we're going to be chasing a rainbow," said Charles Chappel, a spokesman for COLAB.[23]

Perhaps, but less than three months later the APCD announced that the county appeared to have finally met federal ozone standards for 1991 through 1993. The elation was brief, however, as the county quickly slipped back into noncompliance for the 1994-1996 monitoring period. Air quality in fact had not deteriorated, said Doug Allard, director of the Air Pollution Control District. The county, he said, was the victim of meteorological tricks that had concentrated air pollution so frequently at one place that it pushed the county into noncompliance under the federal monitoring formula. Air quality standards also had become more stringent.

Allard and his predecessors—John English and Jim Ryerson—worked under severe pressures and handicaps. The oil industry, the Interior Department, and an often unfriendly governing board bitterly resisted regulations that the Air Pollution Control District felt were necessary to rehabilitate the county's atmosphere.

Meanwhile, the EPA took over air quality planning in Ventura County—often cited as more business-friendly—because that county's efforts were regarded as less than good faith. And the reorganization that merged—or submerged—Santa Barbara County's Air Pollution Control District with two other county agencies was undone within a year. State legislation restored the district as a separate agency, and gave all the county's cities representation on the governing board.

Secretary of the Interior Walter Hickel met with a California delegation in Washington in May 1970 to discuss greater environmental protection for the Santa Barbara Channel. Lois Sidenberg, at far right, was a longtime president of Get Out Oil. (Department of the Interior photo)

Chapter Thirteen

The Distress of Transporting Oil

Producing oil is one thing; getting it to a refinery is something else, especially when the oil is thick as molasses, the refineries are far away, and the producers and public agencies cannot agree on how to transport it. Most of California's offshore oil went to Los Angeles or San Francisco areas. Some also was transported to Oregon, Washington, or Texas. The only refinery in the three counties along the Santa Barbara Channel or the Santa Maria Basin was Unocal's in San Luis Obispo County, and it had no excess capacity to share with other producers. The many nearby processing plants did not refine crude oil into gasoline, diesel oil, or other consumer goods. They did little more than separate water and some liquid natural gases from the crude.

Producers preferred to ship oil in marine tankers which usually charged less than pipeline companies. Moreover, tankers could vary their destinations; pipeline routes were fixed. Oil companies that owned tankers preferred to pay shipping fees to themselves rather than to pipeline companies.

California's coastal communities wanted as little as possible to do with oil tankers, notorious air polluters[1] and maritime hazards. They interfered with commercial fishing and carried the constant threat of catastrophic oil spills. The threat, even without the spills, aroused the opposition of the tourist industry, an important business along the California coast. Knowing what coastal oil spills do to marine resources had made environmentalists of many who would not otherwise be concerned. In the sixteen years from 1968 through 1982 there was an average of nearly one major tanker accident every other week somewhere in the world.[2] Throughout the 1980s, tanker spills in American waters alone averaged about 15,000 metric tons of oil per year, ranging from 6,500 tons in 1987 to 40,000 tons in 1989. Tanker and barge spills, large and small, averaged about 800 per year.[3]

Coastal Californians fretted over many accidents and near-misses. Some were a series of mishaps involving a vessel named *Sansinena*. Just before dawn on September 19, 1973, the tanker *Sansinena* was leaving the Santa Barbara Channel at the narrow end near Anacapa Island. It rammed the fishing vessel *Fort Pioneer* which—bailing all the way—limped to Ventura with ten people aboard. The tanker, unscathed, continued to Los Angeles where two years later it blew up with a blast that, according to writer Eric Nalder, shook buildings forty miles away.[4] Union Oil tempted fate and named a new tanker *Sansinena II*. In March 1978, passing through the Santa Barbara Channel after dark, it smashed into another fishing boat, the fifty-eight-foot *Chelan*. This time the halibut trawler went to the bottom before the Coast Guard arrived on the scene. The *Sansinena II* crew rescued the seven people aboard.

Little or no oil was spilled in these tanker encounters with smaller vessels, but collisions of even non-tankers—dry cargo carriers—can spill a significant amount of oil and other ocean pollutants. On December 29, 1973 the United States Navy cargo ship *Pvt. Joseph Merril* and the freighter *Pearl Venture* rammed each other off Morro Bay. The spill of about 16,000 gallons of oil was not considered catastrophic, but Get Oil Out of Santa Barbara maintained the accident illustrated the impossibility of containing and retrieving oil on rough waters.

Another non-tanker accident dramatized the marine hazards along the coast of California. As the sun rose on September 21, 1987, the cargo vessels *PacBaroness* and *Atlantic Wing*, slammed into each other off Point Conception. As described in Chapter Twelve, the 564-foot *PacBaroness* went down.

There were tanker accidents, too, in the neighborhood. In January 1971 two Standard Oil tankers, the *Arizona Standard* and the *Oregon Standard*, rammed each other just off San Francisco Bay and spread a heavy layer of crude on the water. On October 31, 1984, the tanker *Puerto Rican* exploded off San Francisco and broke in half three days later, disgorging 25,000 to 35,000 barrels of oil into the coastal waters. There were also some close calls.

A few weeks later a barge with 4,500 barrels of oil-water mixture drifted out of control for sixteen hours in high seas and strong winds in the Santa Barbara Channel before it was secured and brought under control. No oil was spilled. On April 19, 1984, the *Sealift Pacific*, a 587-foot tanker, was carrying six million gallons of diesel oil southbound several miles off the Monterey County coast when it lost power. The skipper decided to let the vessel drift shoreward until it reached waters shallow enough to drop anchor in an attempt to hold the tanker off the beach. When the ship was a mile and a half offshore in 240 feet of water, one anchor was dropped. In unusually calm seas, the anchor gripped the bottom and held. The vessel was stopped short of grounding in the only range of the southern sea otter.

The *Puerto Rican* demonstrated that a tanker need not run aground nor collide to make a mess. More freak examples were to come. In November 1988 the propeller of the tanker *Chevron Mississippi* clipped an underwater hose carrying crude oil to a pipeline off Morro Bay. It was capped after about one hundred gallons spilled. On February 7, 1990, the crew of the tanker *American Trader* guessed wrong about the water depth while mooring off Huntington Beach. It punctured the hull as it ran over its own anchor in shallow water. Nearly 400,000 gallons of crude poured out. A little more than a year later and not far away, the tanker *OMI Dynachem* snagged its anchor on an underwater pipeline off El Segundo in Santa Monica Bay. About 21,000 gallons of petroleum distillate escaped.

Industry boasts of the great amount of oil transported *without* spillage. But a small percentage of that amount, when spilled, is suddenly a lot of oil. And as Nalder writes in *Tankers Full of Trouble*, the industry was not doing its best to prevent it.

Small wonder, then, that Santa Barbara and other coastal communities preferred pipeline transport. Oil from abroad or from Alaska had to come in tankers. But oil produced in California's coastal waters could go all the way to refineries in pipelines, and that was obviously the popular—if not the industry's—choice. Pipelines would eliminate tanker loading, and reduce the amount of tanker offloading, both of which were gross air-polluting operations. They would also reduce the risk of tankers maneuvering in the close quarters of harbors. With advanced leak detection and automatic shutoff systems, pipeline technology appeared to be advancing faster than tanker safety design.

In 1984 Santa Barbara County saw an opportunity—and felt a public demand—to control tanker-loading in the Santa Barbara

Channel. While old tanker terminals were being phased out, and before new ones could be applied for, it ruled that:

> If an onshore pipeline for transporting crude oil to refineries is determined to be technically and economically feasible, proposals for expansion, modification, or construction of new coastal dependent oil and gas processing facilities shall be conditioned to require transportation of oil through the pipeline when constructed, unless such condition would not be feasible for a particular shipper.

Significantly, "feasibility" was to be determined by the county, not by an oil company. The question of who determines feasibility—the California Coastal Commission or the corporation—had delayed the Exxon project in the channel for a decade.

Despite public apprehension over tanker casualties and despite the enactment of the county's pipeline regulations, the Arguello Producers, usually represented by Chevron, repeatedly sought approval to move much of its oil from Santa Barbara County by tankers. Starting late in 1987, the consortium of thirteen oil companies with three huge production platforms off Point Arguello and a gigantic processing plant in Gaviota, applied five times for tanker permits. For seven years, the oil companies, the county, the California Coastal Commission, the California Lands Commission, commercial fishers and irate environmentalists struggled over the issue. The Arguello Producers agreed to using pipelines eventually, but they wanted permission to ship by tankers in the meantime.

But as the years wore on, the county and the Coastal Commission lost faith in the company's assurances that it would ever build or use pipelines, and conditions for interim tankering became increasingly stringent. Since it could not get the kind of tankering permit it wanted from the county and the state, Chevron sued both governments for damages suffered because production was being held to less than production capacity.

Governor Wilson's office was concerned about the loss of oil revenues, especially during a serious economic slump, and in January 1992 summoned a group to try to resolve the dispute through a consensus tankering agreement. Among the parties invited to participate was a Santa Barbara coalition of five environmental groups. No agreement emerged from the settlement attempt, but the environmental coalition now had a firm grasp of the issues and a collegial working relationship. It remained organized as an effective counter-influence to what they considered unwise offshore oil development.[5] Many other groups cooperated with the core coalition, creating an even stronger voice.

Chevron and its partners made a few tanker runs in 1993, but lost their permit when they failed to meet requirements indicating progress toward pipeline arrangements. Meanwhile, Arguello Producers' oil went through an interim arrangement of pipelines running from Gaviota eastward to Kern County and then south to the Los Angeles area.

The Arguello Producers entered 1995 without a tankering permit, but by mid-1994 produced an average of 75,000 barrels a day which moved out in whatever pipeline capacity was available. Daily gas production was about 22 million cubic feet.

Public opposition to tankers led to increasing hostility toward oil companies as they resorted to actions widely regarded as bad faith. For several years Exxon circumvented state and local policies by anchoring a processing plant and tanker-loading terminal just beyond California's three-mile offshore jurisdiction.

Even after it built its processing plant onshore and agreed to abide by the county's pipeline policies, Exxon devised a new way to circumvent those requirements. All of Exxon's crude left the county by pipeline, but some of it went 300 miles north to Martinez where tankers were waiting to take it back south 400 miles along the California coast to Los Angeles refineries. Thus, oil was shipped 700 miles to refineries that were only 100 miles away. Exxon explained that there was not enough pipeline capacity available from Santa Barbara to Los Angeles. That satisfied the Minerals Management Service and the California Coastal Commission which, under Exxon's threat of another lawsuit, agreed with tankering from Martinez. It did not satisfy Santa Barbara County whose officials pointed out that Exxon had agreed in 1987 to ship via pipeline from the Santa Barbara Channel *all the way to refineries*, not just to a trans-shipment point.

The county saw little benefit in ending tanker-loading in the channel, only to have tankers move downcoast past the county. Exxon agreed to keep its tankers fifty miles off the mainland, but the single-hulled vessels would be only about fifteen miles off the Santa Barbara Channel Islands National Park, and even closer to the National Marine Sanctuary which surrounds the park. En route, the tankers passed the Monterey Bay National Marine Sanctuary, Los Padres National Forest, the Big Sur coast, and the southern sea otter range.

Exxon could have requested from Santa Barbara County an interim tankering permit, but insisted it did not need such permission to load and operate tankers beyond the county line. The county saw this as a new interpretation by the company eight years after it had agreed to the county's oil transportation policies.

Now the county stood alone again to cope with the defiant corporation which went to court to avoid complying with a county permit. Without support from the Interior Department and even the usually-helpful California Coastal Commission, with court costs mounting, and with the possibility that it might have to pay Exxon's costs as well, the county agreed to settle. Exxon gave no ground on the tankering issue, but in 1996-1997 it shipped little or no oil from Northern California. The jurisdictional issue was moot for the moment.

But there was more than just the jurisdictional matter. The Sierra Club and the Santa Barbara County Citizens Planning Association, represented by the Environmenetal Defense Center, went to court to seek a judicial declaration that the Interior Department had not complied with the National Environmental Protection Act and other federal laws. The environmental groups asserted that by approving tankering, the Interior Department had substantially changed Exxon's federally-approved development plan without notice, without hearings, and without consultation with either the state or Santa Barbara County.

The Interior action was seen as too important to let stand as a precedent. In July 1997 the plaintiffs agreed to a settlement based on an Interior letter to Exxon that lifted Interior's permission to tanker Santa Barbara oil.

Chapter Fourteen

Relics Removed For A New Generation of Technology

As vast new oil industrial enterprises cropped up along Santa Barbara County's coast, onshore relics of earlier endeavors were slowly vanishing. Many created costly county and statewide problems. At Summerland, where it all began, state and federal agencies were still struggling a century later to plug leaking wells and to remove oil industry debris from the world's first offshore operations. (See Chapter One.) "Some 412 improperly plugged wells could exist in the Summerland area," the county reported.[1]

After World War Two, oil companies installed oil and gas processing plants, oil storage tanks, marine terminals, and other industrial components in canyons along the Santa Barbara Channel. Before the end of the century, these plants had served their purpose. Some were shut down, others limped along. Viewing these hulks as remnants of a bygone era, the county began in 1989 to phase them out. Industrial zoning was revoked, and the land was rezoned for homes, recreation, or other non-industrial uses. Oil became a legal non-conforming enterprise; the companies still operating could stay in business, but their plants could not be enlarged, replaced, upgraded, or remodeled.

Gradually—and often reluctantly—the companies removed their plants and equipment, but something remained. Operating with almost no public oversight, much of the industrial processes heavily contaminated the soil and streams with oil, PCBs, and other toxins and pollutants. The damage usually went undetected until after operations were terminated. The results of poor regulation "range from highly contaminated lands to the presence of facilities that ceased operations years ago but remain as rusting eyesores and impediments to sound land use," the county reported in 1994.[2] Rules for proper abandonment were missing from the old operating permits, and the state and county sought to impose cleanup requirements after a generation of land, water, and air degradation.

Probably the most serious contamination was at the site of Shell Oil's gas processing plant on the coast twenty-five miles west of Santa Barbara. The soil was so deeply and extensively permeated by PCBs and other pollutants that it took ten years to design an acceptable cleanup plan after the environmental damage was discovered in 1986. Before the site was restored and revegetated, about 8,500 cubic yards of tainted soil were dug out and hauled away.

While these ancient operations were being phased out, the new offshore oil industry settled into two consolidated sites where environmental monitors kept watch. Each company now knew it must run a neat operation and leave a clean, restored setting when it was over.

In light of the devastation wrought by earlier offshore oil activity, and the enormous pressure for a new generation of oil industry construction, Santa Barbara County in the last quarter of the century had done remarkably well in cleaning up and protecting its coastal resources.

Huntington Beach in Orange County also had a cleanup problem. Seventy-one oil derricks sprouted on its long oceanfront when only few people lived there. As the city grew it wanted the derricks off the waterfront, but without a hostile confrontation with the industry. Oil was a big part of the economy of the city which became the fastest-growing municipality in the nation, erupting from 11,000 in 1960 to 170,000 in 1980. The city council adopted incentives to remove the derricks from the beach. It rezoned parcels for oil drilling inland of the coastal highway, and required oil companies to unitize operations in coalition drilling. Allowing each corporation to drill its own cluster of wells would have even more heavily inundated the city in wellheads.

Inland of the highway, the city set aside zones for oil drilling—current and future—while protecting residential neighborhoods. Some plush homes were moved—not demolished—in the process. By the early 1990s, the beach derricks were gone and waterfront pipelines buried, replaced by lifeguard stations and snack shops along a wide sandy beach.

Off Santa Barbara County, the new was replacing the old. In 1974 Phillips Petroleum gave up on Platform Harry which it installed in 1961 about one and a half miles south of Point Conception. A converted drilling barge to which legs had been added, the platform was relatively easy to remove. With its legs cut off, it floated free and was towed away. At about the same time Texaco shut down two tidelands platforms, also in the west end of the Santa Barbara Channel. Platform Herman off Point Conception and Platform Helen west of Gaviota had produced at modest rates since the early 1960s, and now production was no longer paying its way. The platforms stood idle for several years before they were removed in 1988.

These three platforms were far from urban areas, and many Santa Barbarans did not even know they were there. Four conspicuous platforms off Summerland and Carpinteria, installed in the late 1950s and 1960s, were shut down in 1993 and removed in 1996. That left only Platform Holly in the Santa Barbara Channel's state waters, and it had only a few years left. State and federal agencies required that when platforms are abandoned they must be cut off below the ocean bottom, leaving no trace that they had ever been there. But there was constant pressure from sports fishing groups, some segments of the oil industry and some public agencies to leave part of the platforms under water as fishing reefs.

Even wellheads out of sight on the sea floor were being pulled out. In the late 1990s, six companies with oceanbottom wellheads in the Santa Barbara Channel began removing eighteen subsea wells that were installed in the early 1960s between Point Conception and Summerland.

Much of the ancient oil industrial complex, which by the 1990s could only be called junk, was being removed. Platforms installed in the 1980s and 1990s were by no means replicas of those being displaced. The new structures were far larger and in much deeper water. The enormous production machinery, once powered by air-polluting internal combustion engines, now ran on electricity via cables from shore. The new technology was bigger, and in that sense more intimidating. But it was operating under closer scrutiny of governmental agencies and a more sophisticated and activist public.

Exxon, which had been operating one platform and shipping its production in tankers, installed two more platforms in the Santa Barbara Channel, completed its onshore oil processing plant at Las Flores Canyon, and shipped oil via pipelines. Some of it, however, was transshipped from Northern to Southern California by tanker. (See Chapter Thirteen.)

To handle Exxon's gas production, Pacific Offshore Pipeline Company doubled its Las Flores Canyon gas processing plant capacity to 60 million cubic feet per day.

Exxon oil production peaked at just under its permitted maximum of 140,000 barrels a day, and by 1997 it had dropped to about 85,000 barrels. Oil field production customarily peaks early and then begins a prolonged decline.

But the Chevron Point Arguello Project, with three platforms in the far west end of the channel, was a disappointment from the start. It peaked in mid-1993 at 89,000 barrels a day, less than half what the producers had anticipated. By 1997 it dropped to about 30,000 barrels daily. The field, which Chevron in the early 1980s estimated at 350 million to 450 million barrels of recoverable oil, later was found to be closer to 200 million barrels. Given this turn of events, Chevron indicated the producers planned to abandon the field around the turn of the century, many years earlier than originally intended. The crude went to Los Angeles, mostly by pipeline, and the Southern California Gas Company bought the gas, coming up at about 10 million cubic feet per day.

In 1997, twenty-seven platforms and six islands were in production off Southern and Central California. After four platforms in state waters were removed in 1996, twenty platforms and one production island remained in the Santa Barbara Channel and the Santa Maria Basin.[3] (See Appendices E and F.) Altogether, they were producing 250,000 barrels of oil per day, about four percent of total American production. About 75 percent of that came from production[4] in the federal Outer Continental Shelf.

In state waters, the greatest oil producer was the Wilmington field off Long Beach with nearly 44,000 barrels a day in 1996. That was 75 percent of all state offshore output.

Offshore oil activity appeared to have passed its peak. President Bush had cancelled further federal leasing off California until after the turn of the century—except in the Santa Barbara area where he said leasing could resume in 1996. The Clinton administration cancelled any leasing in the Santa Barbara Channel until at least 2002. In Sacramento, Governor Pete Wilson in 1994 signed a bill drafted by Assemblyman Jack O'Connell of Carpinteria that permanently banned oil activity in all state waters that were not already leased.

The Lull—Just Intermission?

Despite these actions, the Interior Department and the California Lands Commission were restive. At the beginning of 1994 Interior indicated it had no plans for more leasing off California. During that year Interior secretary Bruce Babbitt, on a campaign visit to Santa Barbara, intoned: "No more drilling!" off California's shores. Even as he spoke, his department was drafting the nation's offshore leasing schedule that would run through 2002, and California was still under study. In the draft report published in July 1995, however, the Interior Department acknowledged the intensity of California's protest, and gave up. According to the draft report:

> Many [Pacific] coastal residents remain strongly opposed to any increase in oil and gas activities....[T]he State has formally opposed any leasing off California. Local government policies have reflected this opposition as well.[5]

The final five-year leasing report in 1996 reiterated Governor Pete Wilson's protest against any new leasing, and added that the California Coastal Commission felt the same way.[6]

While the Interior Department had abandoned hope of new leasing off California for five years, it was still pressing for action on old leases not yet in production. It acknowledged that local governments were resolute against more leasing, but in the 1995 draft report the department maintained it found "recent cooperation between the MMS Pacific Office and officials from some local counties in California [that] has helped to identify key concerns, to pave the way for increased production from existing leases, and to lay the groundwork for possible agreement on the conditions under which limited lease sales might be held in the future."

Washington had clearly not given up on more California offshore oil activity. In the early 1990s the Interior Department awarded millions of dollars, much of it to the University of California at Santa Barbara, for several studies that would indicate the impacts of—and public response to—future leasing and production.

The Interior Department strongly implied that the public did not understand the value of such activity, and that opponents were being misled by obstructionists. In an April 26, 1993 memo, MMS indicated it would fund a study that "will assist MMS and other agencies in the decision-making process for selecting potential areas for leasing."[7] The memo to prospective bidders for the contract stated:

> ...many proposed OCS oil and gas developments of the 1980s have met with sufficient community opposition to stifle or delay proposed developments. Many local citizens have not been convinced by the technical, scientific, and economic arguments that OCS development is in their communities' or the nation's best interests. With the important and vocal role the public plays in the OCS development, future projects may also be jeopardized by an increasingly organized and sophisticated opposition. As development of OCS resources is predicated in large part on community acceptance, an understanding of the social context (both historic and contemporary), and the role of the local petroleum industry in that context, is critical.

The next year MMS requested another study, this one concerned with ways to activate the unexploited leases off San Luis Obispo, Santa Barbara, and Ventura counties. This $1.5 million study was to:

> ...analyze the onshore constraints associated with the development of existing undeveloped offshore oil and gas leases....The concerns about the onshore constraints and possible degree and pace of exploration and development of these leases have been communicated to the Minerals Management Service and the State Lands Commission in both oral and written responses to past lease sales and plans of exploration and development.[8]

Even without more lease sales, about fifty existing federal leases and many state leases had not yet been tapped. Oil companies were discouraged by the low price of oil and by the reluctance of coastal counties to make land available for the industry. The MMS diligently sought ways to stimulate activity in the idle federal tracts. J. Lisle Reed, now head of the MMS's Pacific region, organized a coalition of local officials in the tricounties to consider how to place the undeveloped leases into production.

It appeared that Interior was facing a problem of its own making; finding more onshore space—or local approval—for the offshore industry. During the leasing rush of the late 1970s and early 1980s, many private citizens and public officials had warned that immense lease sales were coming faster than the coastal communities could cope with them. Then came the series of local initiatives in the late 1980s that prohibited or hindered construction of onshore support facilities for offshore oil extraction.

This stern public resistance was documented in the first MMS-funded study by sociologists Harvey Molotch of UC Santa Barbara and William Freudenburg of the University of Wisconsin. In three volumes they demonstrated a steadfast or increasing resistance to additional oil operations along the coasts of Santa Barbara, San Luis Obispo, and Ventura counties. In the latter, the oil industry no longer had the public support it once enjoyed, and in the other two the historic opposition was at least as firm as ever.[9]

On the state level, the legislature's ban on more leasing did not prohibit drilling in existing leases. Two companies, in fact, had plans for tapping into the Santa Barbara Channel under state waters by drilling from onshore. Directional drilling was nothing new, but the technology was now so refined that it was regarded as different, not just an improvement of the old technology. Drilling could now be directed almost horizontally once the drill bit completed a sweeping arc at the desired depth. And it could reach out several miles. A sprawling oil reservoir could be tapped more effectively by a horizontal well since most oil deposits were broader than they were deep.

The prospect of drilling into offshore deposits from onshore seemed attractive to Charles Warren, executive officer of the State Lands Commission, because offshore oil could be produced without an offshore platform. Many Santa Barbarans said this merely shifted drilling from offshore platforms to onshore locations, intensifying the degradation where people live. But Warren encouraged Mobil to use this technique to penetrate the Ellwood offshore field. The company wanted to drill onshore wells near the UCSB campus, homes, schools, and coastal wetlands. But UCSB bought the property in 1995 for faculty housing, and denied Mobil permission to drill on university land.[10]

Molino Energy Company, meanwhile, had plans to drill an onshore gas well with this technology twenty-eight miles west of Santa Barbara, two miles from the Gaviota processing plant. Measure A on the county's March 1996 ballot would prohibit drilling there without approval of the county's voters. When the measure qualified for the ballot, and even before the election, Molino moved its proposed operation to the Gaviota oil industry parcel that was set aside for such activity.

This move split the Santa Barbara Environmental Coalition. Molino agreed to conditions requested by coalition leaders, but Get Oil Out maintained drilling would imperil rare plants already on the endangered list, the rigs would violate county height restrictions, and there was a strong suspicion that the gas-only operation would soon incorporate oil drilling as well.

Sure enough, in April 1997 Benton Oil & Gas of nearby Carpinteria indicated it was buying a forty percent interest in the Molino project and intended to go for oil as well as gas. This would require additional permits, but many feared the momentum was there for approval.

The struggle over offshore oil development off California obviously would continue into another century.

Chapter Fifteen

After A Century, What Did It Amount To?

Whether all the offshore California oil production amounted to much depends on what you compare it with. In 1996, it totalled 84.1 million barrels, a lot of oil until you consider that it was about two percent of the nation's rate of consumption. State offshore production peaked in 1969, just as federal Outer Continental Shelf production was beginning. Production from state waters had more than a two-thirds' century head start, but the Interior Department and the deepwater oil industry, starting with 2 million barrels in 1968, were up to 25 million two years later. Oil production in federal waters off California in 1996 was 64.1 million barrels, more than three times the 20 million barrels produced in state waters, and increasing steadily. Total oil production from offshore California through 1996 was about three billion barrels.

Similarly, natural gas output in state waters peaked about the time the industry began exploiting the federal waters. State production reached 85.7 billion cubic feet in 1968, just as production was beginning in federal leases. Output in federal waters reached 52.5 billion cubic feet in 1985 before it began tapering off. In the 1990s, federal gas production turned up again as more platforms were added to the Point Arguello and Santa Ynez units off Santa Barbara County. In 1995 California offshore gas production was 47.5 billion cubic feet, nearly ninety percent of which was from federal leases. Total gas production offshore California through 1995 was about 1.8 *trillion* cubic feet.

With all this production, there should have been considerable public revenue. And there was. The federal government collected nearly $90.7 million in 1993 from California offshore oil and gas production. This included rent on the lease tracts, and royalties, which are fees assessed on the volume of production. The federal government also received the money bid on the leases, but there had been no California offshore lease sales since 1984. The best year by far was 1981 when the Interior Department received more than $2 billion from bids in lease sale 53 off Central California. Since the federal government began leasing off California in the 1960s, it had collected nearly $5.5 billion in public revenue from Pacific oil and gas operations.

The state, too, collects revenues for oil and gas operations in the tidelands. With production far past its peak and no new leasing, state revenues from this source were meager. In FY1996-97, California collected about $91.5 million from offshore oil and gas royalties. When leases were sold and production was flourishing, of course, public revenues were pie in the sky.

With all this money available to public coffers, why had California communities resisted the coastal industry? Most public revenue was collected by Washington and the state of California. Local governments got little except the privilege of providing the services and resources demanded by the offshore oil industry.

To put it gently: "Revenues from offshore Outer Continental Shelf development do not normally flow back to local jurisdictions in a manner commensurate with impacts." That is from Michael Powers of the Santa Barbara County staff who was assigned the task of inducing the oil companies to help pay for the county facilities, resources and services required by the industry. "While there may be direct, tangible national and state benefits due to resource development, the local benefits are very limited," Powers said.[1]

Local governments, of course, can tax properties within their borders, and use the revenue to provide the usual services. But the equation fails when a major part of a mammoth industrial plant is outside the community's borders and beyond the reach of the tax assessor. Local governments can tax oil and gas operations in the tidelands, but the federal waters are out of bounds. For several years Exxon produced, treated, and shipped oil from Platform Hondo off Santa Barbara County beyond the county's ability to impose air quality regulations or collect taxes. The natural gas came ashore for processing and distribution via the POPCO gas plant which was taxed by the county.

But, as we have seen, county property tax revenue fell short of covering the cost of local governmental services required by the industry. Companies sometimes boasted that their projects, if approved, would greatly increase the county tax base, but the results were often disappointing. After prolonged appeals, the companies paid much less than the initial assessment. Annual tax revenues in Santa Barbara County often remained uncertain for years while oil companies appealed their tax assessments.

Sometimes the appeals stemmed from honest disagreement between the county and the companies, sometimes it was a strategic industry ploy, or a gross error in a company's initial evaluation of an oil field potential. When a tax was appealed, the county collected the full assessment, then impounded part of it until the issue was settled. In 1997, after hearings before the County Assessment Appeals Board, the county returned $1.8 million to Chevron and its partners in the Point Arguello Project. Having impounded $6.3 million during a four-year dispute, the county could afford the settlement, or "stipulation," as the Assessor's Office preferred to call it. But schools and other special districts had to wait for their share of the revenue while the matter was in dispute.

In this case the industry partners had grossly overestimated the scope of the Point Arguello field. Estimates fell from the initial 350-450 million barrels in the 1980s, to about 200 million. Instead of a thirty-year project, it became a fifteen-year endeavor, to be abandoned shortly after the turn of the century. Under the formula used by the assessor's office, this sharply reduced the assessed valuation of the property.

Schools and other agencies and districts not only had to wait for their share of the revenue, but it became clear that it would not continue as long as once hoped. Ken Pettit, Santa Barbara County clerk-assessor, and his aide, Richard Holly, said this was a renewed warning to public agencies not to rely heavily on boom-bust industries.

California cities and counties had the additional taxing constraints imposed by two ballot initiatives. Proposition 13 limited the assessments on which taxes could be imposed, and Proposition 4 limited the amount a local government could spend no matter how much it collected in taxes. And as the state's economy deteriorated in the 1990s, Sacramento raided local coffers to pay the state's bills.

It is not that the industry was not spending money and paying taxes in the tricounties. In 1990 it spent more than $100 million in the area, and a little more than half of it "stuck." That is, it was spent in the counties. In addition the companies paid more than

$1.8 million in local taxes, most of it in Santa Barbara County. But the county faced three oil production projects, major pipeline construction and a tanker terminal proposal in the 1980s, none of which could be accommodated with normal county revenues. Special fees would have to be charged to cope with "socioeconomic" costs which varied widely in the impact reports for the several projects.

Socioeconomic impacts included the need for affordable housing for workers and their families, and additional public services: water, roads, electricity, social services, waste disposal, fire and police protection, and, of course, schools. The county had to know—in advance or as the demands arose—how many people would move in, where they would live, their demand for public services, and other stresses of industrial growth. A Socioeconomic Monitoring and Mitigation Program was developed for San Luis Obispo, Ventura and Santa Barbara Counties, although San Luis Obispo dropped out early because the impacts were not serious there. The oil companies did not like some of the counties' assumptions, but agreed that they were better than those in environmental impact reports which, on this issue, were weak and contradictory.

The system was put to use in 1986, and during the first seven years the companies paid $5.9 million to Santa Barbara County, sixty-five percent of it to pay for housing. Ventura County received $1.9 million.

The plan was described as an "insurance policy" for both the communities and the industry.[2] With the impacts reviewed annually, the system protected both sides from relying on an initial set of unreliable forecasts that could later inflict unnecessary costs on one side or the other.

That helped. But it was not enough. There were still "residual environmental impacts to coastal resources" for which there was little or no compensation from the oil industry. To pay for the "residual impacts," Santa Barbara County imposed a special fee crafted to minimize damage to coastal recreation, visual aesthetics, tourism, and coastal resources. "Where such impacts cannot be mitigated entirely through direct measures, the fund offsets the impacts by enhancing the coastal resource at another location or in another way."[3]

The fund was the Coastal Resource Enhancement Fund, and the first assessment against four major oil companies in 1988 was more than $2 million. After that the annual fees ranged from $660,000 to $907,000. The money was used to buy and manage coastal lands, and to build parking lots, educational exhibits, conservation projects, and recreational accommodations.

The cost of producing oil along the Santa Barbara County coast was enormous, both to the industry and to the county. The industry required county permission to do business, and was compelled to pay dearly for it. Their permits were drafted after detailed review by the county's Energy Division and Air Pollution Control District whose budgets came largely from industry fees.

In an attempt to help communities cope with these costs, the California legislature in 1966 required the state to provide the counties twenty percent of oil and gas royalties collected on any new tidelands production projects. It was unlikely to help much. Since new tidelands leasing was now prohibited, only future production programs on existing leases would be affected, and there did not appear to be much of that in the offing.

But if future royalty revenue was bleak, the state and federal governments were looking also to the past for offshore oil royalties on which they said they had been short-changed. Although the Interior Department had long believed that the industry was underpaying royalties, only in 1996 did it file a claim for $440 million from eight major and several minor oil companies. This, the government maintained, represented underpayment of royalties from oil production on federal property in California and off its coast since 1980. The companies were accused of paying royalties on their "posted" price, while selling it at a higher price.

More aggressive than the federal government, California in 1992 collected more than $350 million from seven major companies for onshore and offshore oil royalty shortfalls. That came in a settlement, indicating that the state felt it was less than a full figure.

After a century of experience with offshore oil activity, the state concluded that it had had enough. Governor Pete Wilson and his Resources Agency concluded in a 1997 report:[4]

> Future oil and gas leasing off the California coast would likely cause unacceptable adverse impacts to offshore resources and coastal communities while contributing relatively little to national energy production. A number of factors lead to this conclusion, including visual impacts, navigation risks, drill muds and cuttings disposal practices, air quality impacts, oil spill risks, ecosystem degradation, and uncertain cumulative impacts from existing, approved, proposed, or projected developments.

That was pretty flat out for a state with vocal diversity on almost every other issue.

Coastal Californians throughout the twentieth century were not a monolithic lot. There were commercial fishers and foresters, rugged and independent, in the northern counties of Humboldt and Mendocino; cosmopolitan and progressive-voting residents of the San Francisco Bay area; Central Californians of Monterey, San Luis Obispo and Santa Barbara counties who cherished the beauty and natural resources of their largely-unspoiled coast; coastal rim residents of Ventura and Los Angeles counties where industry and commerce flourished; the resort dwellers of politically-conservative Orange County, and those who enjoyed a strip of coastal amenities alongside heavy urbanization in San Diego County. Eleven hundred miles of diversity.

But no coastal region wanted anything to do with offshore oil. Inland there was less concern, but enough to make offshore oil a vital presidential campaign issue throughout the second half of the century. During that time, however, no California governor except Jerry Brown took a strong position against offshore oil development. Nor did the governors strongly support it. They stepped around it gingerly, and often deferred to other state agencies to defend the oceanfront.

Governor Pete Wilson, who actively opposed offshore oil leasing as mayor of San Diego, as a member of the state legislature, and as a senator, warmed up to offshore oil production for a time after 1991 when he became governor of a state that was short of revenue. Even the 1997 report cited above, while officially representing his administration, was largely the product of the California Resources Agency.

Despite the consensus among coastal residents, there was little united action among the communities. Santa Barbarans and Santa Monicans, for instance, usually did not support each other when either was threatened. San Luis Obispo residents were idle on the issue until oil threatened their shores. Northern Californians, who eventually turned things around in Washington with their overwhelming protests in 1988, were silent on offshore threats until then. With important exceptions, each community fought its battles alone.

But united support was found in Congress where representatives of these communities worked together and wielded an influence far beyond what their numbers would indicate. Environmental lobbyists were active, and to single out one, Richard Charter must be recognized. Representing the OCS Local Government Coordinating Program, he led the battle from 1981 for the annual congressional moratoriums on oil leasing off California.

By withholding leasing funds, Congress checked a serious conflict-of-interest in the Interior Department which carried out the dual roles of fostering and policing offshore oil development. The system drew Interior and the industry into an ever-closer relationship. The industry sought to cultivate a similar relationship with local governments. Instead of proposing industrial projects that complied with carefully-crafted resource protection policies, oil companies often expected local governments to warp and wrench public policies to accommodate corporate industrial plans.

The struggle to maintain the environmental and economic integrity of coastal communities had more champions than could be listed here. Some courageous local government leaders are identified in earlier pages. But because so many public officials were involved in the protest against coastal oil industrialization during the second half of the twentieth century, they have often been just lumped here in a term such as, well, "public officials." Among the outstanding leaders whose great efforts I have not adequately recognized by name were two Santa Barbara County Supervisors: George Clyde in the 1960s and early 1970s, and Bill Wallace who was for twenty years a formidable challenge to the oil industry until his retirement in 1997. Add to them another staunch pair: Gary Hary and Jack O'Connell, both of whom were resolute environmental leaders as they served in the state Assembly and Senate during the tumultuous years.

Far more numerous were the aroused citizens who induced local governments to take courageous action. When that did not work, they enacted laws through ballot initiatives. They got the attention—at least sporadically—of presidents, governors, cabinet members, and legislators.

Struggling against many of the world's largest corporations which possessed seemingly limitless financial resources, they survived on a few dollars and—surprisingly often—prevailed.

Get Oil Out was spontaneous political combustion that was expected to be extinguished before 1970. Instead, it thrived—if one can thrive on austerity—and became a model for local organizations elsewhere with an impressive repository of offshore oil policy information. During its first twenty-five years its presidents included Alvin Weingand, James (Bud) Bottoms, Lois Sidenberg, Frank Sarguis, Steve Boyle, David Anderson, Terry Leftgoff, Henry Feniger, and Joan Kerns. Its leaders traveled far to testify and lobby, even if Lois Sidenberg had to charter planes to get them there. It sponsored public meetings, press conferences, and full-page newspaper ads; collected 200,000 anti-oil signatures, protested presidential appointments, and frequently went to court—sometimes at great financial risk—to hold government and industry to the law. GOO moved its office frequently, each time to lower-rent quarters.

GOO officials appeared regularly in the news, usually with strong, well-founded position statements. They were often crafted by a woman whose name seldom got published. Ellen Sidenberg, daughter-in-law of Lois Sidenberg, provided guidance as executive vice president from its early years to the mid-1980s. As GOO's one (poorly) paid employee, she held the organization together as membership recruiter, fundraiser, archivist, researcher and, often, policy-maker. She had help, of course, but she was the vertebrae to which everything else was attached.

After a decade-and-a-half of exhaustive battles with the oil industry, Ellen was ready to leave it to someone else. In conversations with the press, she said some compromise and accommodation were appropriate. Resisting the inevitable was no longer tenable, she said. But after her retirement, GOO stayed the course: Get Oil Out! The exclamation point was part of the official name.

Two national organizations supported their Santa Barbara units in confronting the offshore oil industry. Armed with irrefutable data based on serious research, the Santa Barbara League of Women Voters dug out embarrassing information and asked disquieting questions. Many heroic workers toiled in the LWV ranks, and I will mention only two: Ruth Saadi, the little retired teacher (with a doctorate), who pulled the mike down to her level and—most respectfully—launched the torpedoes. The other was Connie Hannah, who stepped in when Ruth yielded to time's toll.

The national Sierra Club, too, supported its local representatives such as Fred Eissler and Kalon Kelly in resisting the industrial onslaught.

Although GOO was the landmark group spawned by the 1969 blowouts, two organizations of the same parentage persisted with greater flourish. The Community Environmental Council, nee the January 28 Committee, became a world-recognized environmental training and research institution. Paul Relis and Robert Easton were chief among its founders.

The Environmental Defense Center became a public interest law firm from which many groups in Santa Barbara, Ventura, and San Luis Obispo seek legal assistance and political leadership. Under the guidance of Marc McGinnes, Michael David Cox (later Fox), Phil Seymour, Marc Chytilo, and Linda Krop, among others, it staved off many ill-considered coastal "improvements." Krop became the guru of the Santa Barbara Environmental Coalition, the core nonprofit group concerned with offshore oil activities.

The Santa Barbara County Citizens Planning Association, which antedated the 1969 offshore disasters, constantly had well-prepared people on hand to match the "expertise" of the oil industry and its allies. For many young people, the CPA was a training ground for local government employment. It was always, of course, short of funds. Indeed, most of these groups were always scraping for money, and Selma Rubin, usually sitting on a dozen boards simultaneously, kept them afloat with a flow of personal checks.

Santa Barbara was long the center of resistance to offshore oil development. The community's affection for its privileged environmental setting coincided—and collided—with the profusion of underground petroleum. But thousands of Californians elsewhere eventually became even more militant than Santa Barbarans in protecting their coast which became threatened, but not yet invaded, by the oil industry.

Voters in seventeen coastal cities and counties passed initiatives to prohibit or impede onshore construction related to offshore oil extraction. The governing bodies in ten more communities adopted similar ordinances. One has to applaud Dan Haifley, executive director of Save Our Shores, for organizing the Oregon-to-Mexico campaign for these local actions. Meanwhile, Northern Californians in 1988 turned out by the thousands at the Fort Bragg hearing to sour Washington's appetite for California offshore leasing for the rest of the century.

Those who protested were often labeled anti-business obstructionists. Why are you always *against* everything? they were often asked. Why cannot you be *for* something? Environmentalists sometimes did emphasize the negative, but were often forced to do so. As they defended cherished resources from damaging proposals, they often were characterized as the negative faction. The positive alternative—which Washington and the oil industry resisted—was a new energy policy that would get America off the oil kick; energy sources that would be clean, renewable, safe, and efficient. It was obvious at least as early as 1970 that the combustion of fossil fuels had to end, the sooner the better. But industrial executives and political leaders, with greater funding and access to the media, convinced much of the public that there was no acceptable substitute for gasoline in the tank.

Scientists generally agreed that fossil fuel combustion was adding enough carbon dioxide, a "greenhouse" gas, to the atmosphere to create serious, maybe catastrophic, global warming. Some scientists who questioned this forecast gave government and industry an excuse to delay serious efforts to curb carbon dioxide exhausts. The "soft" energy paths would have been justified because, given a warning, it is not wise to await the results of an anticipated catastrophe before taking action. Moreover, even if the earth's atmospheric temperature would not rise seriously, the combustion of fossil fuel should be ended for another compelling reason—to clean up the air.

Oil does not pollute only when it leaves auto exhaust pipes. The oil industry is a polluter at every step: exploring, producing, processing, transporting, refining, marketing, and consuming. That is even when everything goes right and nothing spills, burns, explodes, or evaporates when it is not supposed to. Beyond that is pollution from the manufacturing and servicing required to run an oil operation.

A new energy policy was what the protestors were *for*, but too often they got trapped into debating the issues as they were framed by the industry and the federal government. They were asked to consider "compromises," such as reducing the scale of a venture, phasing it in, or otherwise softening the impact of the industry. "Compromises," however attractive they seemed, in fact were concessions by the communities at little or no expense to the industry. The oil companies started with nothing but hopes and dreams and something on paper. The community started with its resources and neighborhoods intact. If an oil company cut its plans in half, it compromised nothing because it started with nothing. The community, however, was still asked to offer a substantial amount of its resources—air, water, land, public services, and many amenities—for corporate profit.

As explained earlier, the offshore oil industry paid its way in the community only after the local government devised and carried out complex compensation plans. Customary property taxes did not pay for the community's cost in accommodating the industry. Sandy-beach Californians were accused of enjoying full use of their petroleum-driven autos, but were not willing to have the fuel produced in their communities.[5] They were dismissed as NIMBYs (Not in My Back Yard). But were not NIMBYs often the real "experts" in assessing the consequences of a proposed enterprise? The National Academy of Sciences in 1992 audaciously declared that strong local feelings about things that cannot be neatly quantified or even specifically articulated are as worthy of consideration as the oil industry's contribution to national security and the public treasury. (See Chapter Nine.)

What part this counsel would play in the second century of California offshore oil activity was, of course, yet to be seen.

But whether the crude oil becomes gasoline, diesel oil, or asphalt, each barrel produced will be, in truth, one more for the road.

Remnants of Ellwood Field Piers in 1978. (Robert Sollen photo)

Notes

Chapter 1

[1] Peter J. Fischer, "Natural Gas and Oil Seeps and Geology of the Northern Santa Barbara Basin, California," prepared in 1976 for the Federal Energy Administration.

[2] Hydrocarbon seepage emerges in various densities as natural gas, liquid oil, thicker asphaltum, semi-solid bitumen and sticky tar.

[3] Campbell Grant, "The Carpinteria Tar Pits," *Noticias* (Winter 1962), p. 13.

[4] Bruce W. Miller, *Chumash: A Picture of Their World* (Los Osos, California: Sand River Press, 1988).

[5] Susan F. Hodgson, "Onshore Oil and Gas Seeps in California," a report for the California Division of Oil & Gas, Sacramento, 1980.

[6] Jan Timbrook, anthropologist, Santa Barbara Museum of Natural History, interview, 1992.

[7] Iris Priestaf, "Natural Tar Seeps and Asphalt Deposits of Santa Barbara County," *California Geology* (August 1979), p. 164. In this reference, she cites Walker Tompkins' 1966 book, *Goleta the Good Land.*

[8] Ralph Arnold, "Geology and Oil Resources of the Summerland District," U. S. Geological Survey *Professional Paper* 321 (1907).

[9] David Banks Rogers, *Prehistoric Man of the Santa Barbara Coast* (Santa Barbara: Santa Barbara Museum of Natural History, 1929), p. 51.

[10] May Lambert, *Growing Up With Summerland 1874-1975* published 1975 by the Carpinteria Valley Historical Society, p. 38. See also David F. Myrick, "Summerland: The First Decade," *Noticias* (Winter 1988), p. 78.

[11] *California Oil, Gas, and Geothermal Resources* (California Division of Oil and Gas, 4th edition, 1988), p. 8.

[12] Myrick, pp. 76-78.

[13] Myrick, p. 99.

[14] Kenny A. Franks and Paul F. Lambert, *Early California Oil: A Photographic History* (College Station: Texas A and M University Press, 1985), p. 20.

[15] Myrick, pp. 95-96.

[16] *California Offshore Oil and Gas Development* (San Francisco: California Coastal Commission, July 1992), p. 2.

[17] Henry Kleine, "Ride With Seaside," *Noticias* (Winter 1988), p. 105.

[18] Stella Haverland Rouse, in a column in the *Santa Barbara News-Press*, n.d.

[19] Rouse, May 5, 1974.

[20] Earnest R. Bartley, *The Tidelands Oil Controversy: A Legal and Historical Analysis* (Austin: University of Texas Press, 1953), pp. 67-69.

[21] Bartley, pp. 67-68.

[22] Charlie Powell, senior petroleum engineer, California Lands Commission, personal interview, Sept. 29, 1991.

[23] Santa Barbara County Energy Div., *Abandonment of Oil and Gas Production Fields and Related Facilities in Santa Barbara County* (draft), 1994, p. 4.

[24] David W. Brown, chief of administrative and information services, California Lands Commission, personal correspondence, Jan. 20, 1993.

[25] Terry R. Bertoline, supervisor of the Santa Barbara field office of the California Lands Commission, personal interview, May 9, 1991.

[26] Quoted by Melinda Burns, "State to plug leaky Summerland oil wells," *Santa Barbara News-Press*, July 9, 1992.

CHAPTER 2

[1] A rotary drill bit is the tool attached to the lower end of the drill pipe; a heavy steel "head" equipped with various types of cutting or grinding teeth, some are fixed, some turn on bearings. A hole in the bottom of the drill permits the flow of drilling mud being pumped down through the drill pipe to wash the cuttings to the surface and also cool and lubricate the bit. *The Illustrated Petroleum Reference Dictionary*, Robert D. Langenkamp, editor (Tulsa: Petroleum Publishing Co., 1980), p. 47.

[2] William Rintoul, *Drilling Ahead* (Santa Cruz:Valley Publishers, 1981), pp., 96-97.

[3] Charles S. Jones, *From the Rio Grande to the Arctic: The Story of the Richfield Oil Corporation* (Norman: University of Oklahoma Press, 1972), pp. 42-43. The detailed story of the early years of the Ellwood field are described in chapters 7, 8 and 9.

[4] Jones, p. 51.

[5] *Draft Environmental Impact Statement for ARCO Exploratory Drilling off Coal Oil Point*, prepared for the State Lands Commission by Atlantis Scientific, June 1980, pp. 2-9.

[6] Variations on the story are in Jones' book, *From the Rio Grande to the Arctic*, p. 53, and in "Oil Conservation in Santa Barbara County," an address by Jones before the Channel City Club in Santa Barbara, Oct. 8, 1956.

[7] *Geology, Petroleum Development, and Seismicity of the Santa Barbara Channel Region, California*, Geological Survey *Professional Paper* 679 (Washington, D. C.: Government Printing Office, 1969), p. 19.

[8] By 1980, a total of 24 oil and gas fields had been discovered off California's coast. Eight are extensions of onshore fields. *California Oil, Gas, and Geothermal Resources: An Introduction* (Sacramento; Division of Oil & Gas. 1988), p. 52.

[9] Bartley, p. 71.

[10] Bartley, p. 72

[11] Bartley, p. 72.

[12] "Survey of Remnants of Oil Producing Facilities on State Coastal Tidelands Between El Capitan Beach State Park and Summerland, California, February-October 1979," prepared by Oceanographic Services Inc., for the California State Lands Commission, January 1980.

[13] The term "tidelands" as used here and by state and federal agencies are the ocean waters extending three miles seaward from the high tide line. Under state and federal law and usage, "lands" includes water areas. Thus, the California State Lands Commission has jurisdiction three miles seaward. And the statute conveying federal jurisdiction offshore is the Outer Continental Shelf Lands Act. The shoreline between high and low tides is the *intertidal* zone.

[14] Much of the information in this chapter is based on Ernest R. Bartley's book, *The Tidelands Oil Controversy: A Legal and Historical Analysis* (Austin: University of Texas Press, 1953), particularly chapters 5 and 7 through 10.

[15] Bartley, p. 136.

[16] T. H. Watkins, *Righteous Pilgrim: The Life and Times of Harold L. Ickes, 1874-1952* (New York: Henry Holt, 1990), pp. 827-33.

[17] Watkins, pp. 827-33.

[18] *U. S. vs. California*, 332 U. S. 19 (1947)

[19] "[C]ourts sometimes appoint a special master to assist in duties that would overwhelm the judges. The Supreme Court uses special masters in cases invoking its original jurisdiction, such as boundary and water disputes between the states. The special master assembles the evidence and writes an advisory opinion or decree, which the Court at its discretion may adopt, modify, or reject." From Jethro K. Lieberman, *The Evolving Constitution: How the Supreme Court Has Ruled on Issues from Abortion to Zoning* (New York: Random House, 1992), p. 497.

[20] This was *not* the same Cunningham who opened the tidelands debate twenty years earlier before a congressional committee.

[21] It is strange enough for federal waters to divide parts of a state, but for certain purposes—including navigation—much of these waters are regarded as international, or "high seas;" California's Strait of Bosphorus.

[22] *U. S. vs. California*, 381 U. S. 139 (1965).

Chapter 3

[1] William E. Kennett, "History of Oil Development in the Santa Barbara Channel," in *Petroleum Geology of Coastal Southern California* (Pacific Section of the American Association of Petroleum Geologists, 1987).

[2] Beth Porter, recorded interview for Oral History Archives of the Carpinteria Valley Museum, August 13, 1982.

[3] Len Swanson, "Sportsmen, Fish Firms Wrathful Over Blasting," *Santa Barbara News-Press*, May 2, 1948.

[4] "This Is It—The Real Battle to Save Beaches from Oil Drilling," *Santa Barbara News-Press*, Aug. 22, 1954.

[5] Kennett.

[6] *Geology, Petroleum Development, and Seismicity of the Santa Barbara Channel Region, California.* Geological Survey *Professional Paper* 679 (Washington, D. C., Government Printing Office, 1969), p. 15.

[7] Although production platforms were something new off Southern California in the 1950s, the industry had experience with such structures in the Gulf of Mexico since 1938. But the waters are much deeper off California's coast. The first Gulf platform, a mile and a half offshore, stood in about 20 feet of water. The first well out of sight of land was 12 miles off Louisiana in 18 feet of water, according to an exhibit in the Unocal Oil Museum in Santa Paula, California.

[8] Californians saw one free-standing production platform as early as 1932. Indian Petroleum Co. held a lease off Rincon in Ventura County and had built piers to reach into the tidelands. About 1,700 feet beyond the longest pier it put up a steel platform in 38 feet of water. Production was meager from its three wells, and in 1940 a storm destroyed the platform. No other California offshore platforms were seen for 17 years.

[9] The THUMS lease was later purchased from the five partners by ARCO.

[10] The Santa Barbara Channel was spared this sale. The U. S. Supreme Court in 1963 had not yet ruled on California's claim to the entire channel as "inland waters."

[11] Harry R. Anderson, assistant secretary; Eugene W. Standley, department's staff engineer; and Donald Solanas, department's regional oil and gas supervisor.

[12] *Santa Barbara News-Press*, Nov. 28, 1967.

[13] George Clyde, "Channel oil: how problem developed," *Santa Barbara News-Press*, Feb. 4, 1973.

[14] Walter L. Healy, "Pro-Oil Groups Outnumber Platform Foes at Hearing," *Santa Barbara News-Press*, Nov. 21, 1967, p. 1.

[15] Robert Easton, *Black Tide: The Santa Barbara Oil Spill and its Consequences* (New York: Delacorte Press, 1972), p. 104.

[16] Walter Healy, *Santa Barbara News-Press*, Feb. 7, 1968.

[17] "Oil Firm Convicted of Polluting L. A. Harbor," *Los Angeles Times*, March 26, 1969, p. 24.

[18] Robert Sollen, "Oilman Says County Doesn't Deserve 'Antibusiness' Tag," *Santa Barbara News-Press,* May 18, 1968, p. A-2.

[19] "Santa Barbara 'Antibusiness'?" editorial, *Santa Barbara News-Press*, May 21, 1968.

[20] Feb. 15, 1968 memorandum from Eugene Standley, Interior Department staff engineer, to J. Cordell Moore, an assistant secretary at Interior.

[21] "Union Oil Begins Building on Base of Channel Platform," *Santa Barbara News-Press*, Sept. 18, 1968. (The article lacks a byline, but the florid prose indicates it was written by someone else on my day off.)

CHAPTER 4

[1] Lee Dye, *Blowout at Platform A: The Crisis that Awakened a Nation* (Garden City, N.Y.: Doubleday, 1971) Robert Easton, *Black Tide: The Santa Barbara Oil Spill and Its Consequences* (New York: Delacorte Press, 1972). Carol E. Steinhart and John S. Steinhart, *Blowout: A Case Study of the Santa Barbara Oil Spill* (North Scituate, Massachusetts: Duxbury Press, 1972). A. E. Keir Nash, Dean E. Mann, and Phil G. Olsen, *Oil Pollution and the Public Interest: A Study of the Santa Barbara Oil Spill* (Berkeley: UC Institute of Governmental Studies,1972).

[2] Steinharts, p. 57

[3] John H. Averill, "Hickel Faces Tough Quiz by Senators," *Los Angeles Times*, Jan. 13, 1969, p. 8, Part I.

[4] Walter J. Hickel, *Who Owns America?* (Englewood Cliffs: Prentice-Hall, 1971), pp. 94-95

[5] Shortly after the Santa Barbara Channel oil well blowout, the Orange County Board of Supervisors asked for a federal buffer zone seaward of the state tidelands sanctuary off Orange County. In June 1970, Orange County, three of its coastal communities and the Coastal Area Protective League hired a lobbyist to promote a bill establishing a federal marine oil-free sanctuary off Orange County. "All future offshore drilling in this area...should be prohibited," The Orange County Board of Supervisors declared in January 1971. These Orange County actions illustrated serious concern outside the immediate crisis zone.

[6] Hickel, p. 99

[7] Hickel, pp. 96-97.

[8] Steinharts, p. 53

[9] Steinharts, p. 68

[10] "Attorney Charges Oil Industry Keeps Experts Out of Lawsuit," *Santa Barbara News-Press,* April 8, 1969.

[11] Easton, pp. 138, 139.

[12] Marvin Levine, "The Santa Barbara Saga—Pre and Post N.E.P.A.," *Lincoln Law Review* (December 1971), pp. 75-76.

[13] Robert Sollen, "When CEQA-NEPA Hit Santa Barbara," *Gildea Resource Center Newsletter*, published by Community Environmental Council of Santa Barbara (Fall 1985), p. 1.

[14] The citation read, in part: "Through his paper he [Stuart S. Taylor] has in this last year exerted not only local but national influence to help preserve our priceless environment for the American people....Mr. Taylor resisted pressures from vested interests and advertisers unhesitatingly at all times."

[15] Noel Greenwood, "Area of Channel Oil Blowout Highly Unstable, Report Says," *Los Angeles Times*, Jan. 9, 1970, p. 22

[16] Hickel, p. 294.

Chapter 5

[1] Lee Dye, "OK of 2 New Offshore Oil Rigs Believed Set," *Los Angeles Times,* Jan. 9, 1971, p. 1, Part II.

[2] Robert H. Sollen, "Morton Denies Permits for 2 Platforms in Channel," *Santa Barbara News-Press*, Sept. 21, 1971.

[3] Bob Sanders, "Seal Beach repeals ocean oil drill ban," *Long Beach Press-Telegram*, June 8, 1971.

[4] Philip Fradkin, "California Coast: No Legislation for Conservation," *Los Angeles Times*, June 19, 1972.

[5] Thomas J. Foley, "U. S. Reverses Stand Barring Oil Drilling Off Santa Barbara," *Los Angeles Times*, Nov. 14, 1973.

[6] Stewart Udall, Charles Conconi, and David Osterhout, *The Energy Balloon* (New York: McGraw-Hill, 1974), p. 96.

[7] *Outer Continental Shelf Oil and Gas Development and the Coastal Zone*, a National Ocean Policy Study staff report prepared for Sen. Magnuson, chairman of the Senate Commerce Committee, Nov. 1974, p. 1.

[8] "Off-California lease tracts disclosed," *Oil and Gas Journal,* July 22, 1974, p. 16.

[9] Robert A. Rosenblatt, "Energy Crisis: Oil Firms, U. S. Caused Shortage," *Los Angeles Times,* July 17, 1973. pp. 1, 3, 18, 19.

[10] Editorial, *Oil and Gas Journal,* Dec. 9, 1974.

[11] Worldwatch Institute, *Energy: The Case for Conservation*, published 1976 under a grant from the Federal Energy Administration. Summarized by Associated Press Feb. 1, 1976. Stanford Research Institute (SRI) study funded by the Federal Energy Administration. Summarized by *Christian Science Monitor* Feb. 19, 1976.

CHAPTER 6

[1] The Minerals Management Service was created in the Interior Department in 1982 to assume management of OCS oil and gas leasing, and supervision of post-leasing activities and functions previously shared by the Geological Survey and the Bureau of Land Management.

[2] Robert H. Sollen, "Exxon plans for offshore facility hit by 11 groups at hearing," *Santa Barbara News-Press*, Oct. 5, 1977.

[3] Summary Minutes, Pacific States Regional Technical Working Group Meeting, March 27, 1981, Los Angeles, p. 1.

[4] Representative John Murphy of New York was subsequently convicted of accepting bribes from federal agents posing as oil-rich Arabs in a federal case known as Abscam. The conviction was, of course, unrelated to the committee report that was signed and subscribed to by eight committee members.

CHAPTER 7

[1] Ed Meagher, "Liquefied Natural Gas—Risk of a Disaster Feared," *Los Angeles Times*, Jan. 12, 1975.

CHAPTER 8

[1] The Santa Barbara Channel Islands National Marine Sanctuary encompasses the four northern Channel Islands, which run parallel to the Santa Barbara and Ventura county shorelines and enclose the waters most sought-after by the oil industry. Also included was Santa Barbara Island further south. Three of the eight islands in the chain which were away from the offshore oil fields, were not included in the sanctuary.

[2] A nautical mile is 1.15 statute miles. It is a minute of latitude, or 6,076 feet.

[3] The Channel Islands National Park, established in 1980, includes all public lands on the same five northern Channel Islands. This includes the islands of Anacapa, San Miguel, and Santa Barbara, and Santa Rosa Island which was acquired later. The Nature Conservancy owns most of Santa Cruz Island, and the National Park Service in 1994 was still negotiating to purchase the remaining 6,000 acres of privately-owned land on the island.

[4] "Watt Ignored Lease Advice, Papers Show," *Los Angeles Times*, May 22, 1981.

[5] Richard D. James, "Controversial Sale of Offshore Oil Leases Tests Reagan Plan to Hasten Exploration," *Wall Street Journal,* May 27, 1981.

[6] Robert Sollen, "U. S. says state oil lease protests are ill-founded," *Santa Barbara News-Press*, April 17, 1981.

[7] Commercial fishers, who protested plans for oil development in northern California shores, were not so concerned about protecting the sea otter off Central California. To them, the voracious otter, whose appearance enthralled many, was a threat to the abalone fishery.

[8] Reoffering sales are comprised of lease tracts that were included in earlier sales, but received no bids or on which all bids were rejected.

[9] Ronald B. Taylor, "Hodel Accuses Coastal Panel of 'Usurping' U. S. Authority," *Los Angeles Times*, June 15, 1987.

CHAPTER 9

[1] Lou Cannon, *President Reagan: The Role of a Lifetime* (New York: Simon & Schuster, 1991), p. 435.

[2] Philip Shabecoff, "Clark Says He Offers a Role to Critics of Sea Oil Leasing," *New York Times*, Jan. 13, 1984.

[3] Resolution of the Orange County Board of Supervisors, March 10, 1987.

[4] Report of the Orange County Environmental Management Agency on the federal 1987-92 five-year offshore leasing plan, February 1987.

[5] Maura Dolan, "Offshore Oil Pact Rejected by Hodel," *Los Angeles Times,* Sept. 11, 1985.

[6] Donald Paul Hodel, "Offshore Oil Production," letter to the editor, *Los Angeles Times*, Sept. 25, 1985.

[7] Associated Press, "Hodel says 'wealthy West Coast environmentalists' stall drilling, *Santa Barbara News-Press*, March 23, 1986.

[8] Miles Corwin, "Angry Residents of State's Northern Coast Jeer Plans for Offshore Drilling," *Los Angeles Times*, Feb. 4, 1988.

[9] William R. Freudenburg and Robert Gramling, *Oil In Troubled Waters: Perceptions, Politics, and the Battle Over Offshore Drilling* (New York: State University of New York Press, 1994), p. 2.

[10] Freudenburg and Gramling, p.4.

[11] Robert Kallman, personal interview, July 2, 1994.

[12] Douglas Jehl, "Reagan Aides Scored in Study of Offshore Oil, *Los Angeles Times*, April 6, 1989.

[13] United Press International, "Wildlife Official Who Raised Environmental Concerns Dismissed," *Los Angeles Times*, Jan. 12, 1989.

[14] William J. Eaton, "Californians Demand Lujan Quit Task Force on Offshore Drilling," *Los Angeles Times*, July 22, 1989.

[15] Interview with Robert Kallman, July 2, 1994.

[16] Outer Continental Shelf Leasing & Development Task Force, *A Report to the President on Lease Sales 91, 95, and 116, Part II*, Jan. 1, 1990.

[17] Jim Little, "Read Our Lips: Locals Tell Bush: No New Drilling," *Santa Barbara Independent*, July 13, 1989.

[18] "U. S. Reveals 'Horrible' Oil Mess," Times-Post Service article in the *San Francisco Chronicle*, April 30, 1969. "Loss to Government Seen in Oil Leases," *Santa Barbara News-Press*, May 2, 1969.

[19] Walter Hickel, news conference in Los Angeles, Nov. 17, 1969.

[20] National Research Council/National Academy of Sciences, *The Adequacy of Environmental Information for Outer Continental Shelf Oil and Gas Decisions: Florida and California* (Washington D. C.: National Academy Press, 1989), pp. 6,7.

[21] Amory B. Lovins and L. Hunter Lovins, "The Energy Saboteurs Are In the White House," *Los Angeles Times*, Jan. 21, 1990.

[22] National Research Council, *Assessment of the U. S. Outer Continental Shelf Environmental Studies Program: III. Social and Economic Studies* (Washington, D. C., National Academy Press, 1992), pp. 90-93.

Chapter 10

[1] Robert Sollen, "Official says state is preparing oil lease off north county coast," *Santa Barbara News-Press*, July 30, 1980.

[2] The Coastal Commission was established first by a vote of the people in 1972, and reauthorized in 1976 by the legislature. The Lands Commission consists of the lieutenant governor and state controller, both of whom are elected officials, and the state finance director, appointed by the governor who was left with only one "vote" among the three.

[3] Correspondence from Robert Hight, executive officer of the California State Lands Commission, Aug. 25, 1994.

Chapter 11

[1] Robert H. Sollen, "EXXON OFFSHORE FACILITIES: EPA planning to impose local and federal air rules," *Santa Barbara News-Press*, (Nov. 23, 1977).

[2] Robert Lindsey, "Oil Industry is Pressing for California Deposits," *New York Times*, (Sept. 7, 1986).

[3] William E. Grant, "County has left no onshore choice to either Exxon or the country," letter to the editor of the *Santa Barbara News-Press*, Jan. 2, 1987.

[4] Keith E. Dalton, "Huffington's hire knows media, ropes," *Santa Barbara News-Press*, July 14, 1994.

[5] National Research Council, *Assessment of the U. S. Outer Continental Shelf Environmental Studies Program: III. Social and Economic Studies* (Washington, D. C.: National Academy Press, 1992), pp. 90-93.

[6] Bob Levy, "Offshore oil foes savor wins in 9 of 9 community elections," *San Jose Mercury News*, Nov. 6, 1986.

[7] Keith G. Hay, "Fish and Offshore Oil Development," *American Petroleum Institute*, 1984, pp. 10-11.

CHAPTER 12

[1] "The Basics of Drilling Fluids," *IMPCO SERVICES* (1979).

[2] National Research Council Panel on Assessment of Fates and Effects of Drilling Fluid and Cuttings in the Marine Environment, *Drilling Discharges in the Marine Environment* (Washington, D. C: National Academy Press, 1983), p. 2. The report stated that about two million metric tons were discharged annually on the entire OCS, and that more than ninety percent of it was in the Gulf of Mexico. The rest was off California, the only other OCS region under development.

[3] Keith Dalton, "Drill mud dumping: Blanket approval opposed," *Santa Barbara News-Press,* Oct. 22, 1985.

[4] NRC panel, p. 2.

[5] Robert Sollen, "EPA regional director backs dumping of waste in channel," *Santa Barbara News-Press*, Dec. 4, 1981, and "Firm Reagan OCS, gas decontrol stances urged," *Oil and Gas Journal,* Dec. 14, 1981.

[6] Personal correspondence, Aug. 25, 1982.

[7] Correspondence from Matt Rodriguez, coastal policy analyst, California Governor's Office of Planning and Research, to Dr. John Costlow, chairman, Panel on Assessment of Fates and Effects of Drilling Fluids and Cuttings in the Marine Environment, Aug. 27, 1982.

[8] Correspondence from Frank Press, chairman of the National Research Council, to John L. Mohr, Sept. 2, 1982.

[9] NRC Panel, p. 4.

[10] News release, American Petroleum Institute, June 20, 1989.

[11] Information packet distributed by Marine Spill Response Corporation, Washington, D. C., March 3, 1993.

[12] National Research Council Committee on Effectiveness of Oil Spill Dispersant, *Using Oil Spill Dispersant on the Sea* (Washington, D. C.: National Academy Press, 1989), pp. 2, 3.

[13] The most immediately obvious impacts of photochemical smog are the yellow-brown haze that cuts visibility, and eye irritation. The most serious health problem is the aggravation of respiratory conditions. Asthma, bronchitis, and other respiratory ailments, as well as cardiovascular disease are aggravated by exposure to ozone. Children and the elderly are most susceptible. Beyond that, photochemical oxidant deteriorates substances such

as rubber, plastic, paint and textiles, and damages agricultural crops and other vegetation.

[14] Robert H. Sollen, "'Outside' sources: air standard threat?" *Santa Barbara News-Press*, Sept. 15, 1978.

[15] Robert Sollen, "EPA chief stifles air quality help," *Santa Barbara News-Press*, Dec. 10, 1982.

[16] Interview with William Master, former assistant director, Santa Barbara County Air Pollution Control District, Nov. 19, 1994.

[17] Interview with Kathy Milway, intergovernmental affairs manager, Santa Barbara County Air Pollution Control District, Nov. 19, 1993.

[18] Marvin Braude, letter to the editor, "Battle for Clean Air Getting Another Blow From Reagan," *Los Angeles Times*, May 7, 1986.

[19] Air Pollution Control District, *1993 Rate-of-Progress Plan: Federal Ozone Standard Countywide* (Santa Barbara County, October 1993), pp. 2-12, 2-13.

[20] APCD, p. EX-2.

[21] In August 1994 an environmentally-oriented majority was restored when the California Court of Appeals ruled that a close and contested election had in fact been won by Bill Wallace, the incumbent. The California Supreme Court declined to hear an appeal. Willy Chamberlin, who had served more than 18 months of the four-year term, was forced to vacate his seat.

[22] Santa Barbara County Air Pollution Control District, *Dispelling the Myths About the Clean Air Program in Santa Barbara County*, On the Air newsletter, May 1992, p. 3.

[23] Nick Welsh, "Clearing the Air the Political Way," *Santa Barbara Independent*, Dec. 2, 1993. Also Melinda Burns, "Group suggests moving ozone monitors," *Santa Barbara News-Press*, Dec. 7, 1993. COLAB later opposed adding another monitoring station on Santa Barbara's "Riviera" where some residents felt that an air pollution "hot spot" was being ignored. The station was installed.

CHAPTER 13

[1] In 1977, the Santa Barbara County Air Pollution Control District estimated that fumes from tanker-loading in the Santa Barbara Channel added 277 tons of air pollution per year, and that it would become worse unless pipeline oil transportation was substituted for tankers. Among the worst offenders were tankers that took on oil from the Chevron processing plant at Carpinteria. But Chevron discontinued tankering there in 1979 and channeled oil from five producing platforms via pipeline downcoast a few miles to the Rincon in Ventura County where it joined with other lines to a Los Angeles area refinery.

[2] Liverpool Underwriters' Association, *Annual Report* (1982).

[3] Mercer Management Consulting analysis of U. S. Coast Guard Oil Spill Database 1981 to 1990.

[4] Eric Nalder, *Tankers Full of Trouble: The Perilous Journey of Alaskan Crude* (New York: Grove Press, 1994) pp. 114-15.

[5] The agencies involved were the California Resources Agency, Coastal Commission staff, Santa Barbara County energy specialists, the Arguello Producers, and the Santa Barbara environmental community. The latter were represented by the Environmental Defense Center, Citizens Planning Association of Santa Barbara County, Get Oil Out (GOO), the League of Women Voters, and the Sierra Club. Henry Feniger of GOO was its forceful and articulate spokesperson during the two months of negotiation. Linda Krop of the Environmental Defense Center became leader, legal advisor and principle spokesperson for the coalition in subsequent appearances.

CHAPTER 14

[1] Santa Barbara County Planning and Development Department, Energy Division, *Abandonment of Oil and Gas Production Fields and Related Facilities in Santa Barbara County*, draft report, 1994, p. 3.

[2] *Ibid*, p. 1

[3] This is a small number of platforms compared with the thousands in the Gulf of Mexico, but each California platform has a far greater impact than those off the Gulf Coast. California's platforms are two to twelve miles offshore, all off urban areas and—with the couple exceptions—visible from shore. Almost all Gulf of Mexico platforms are far offshore out of sight. California's platforms are conspicuous from shore.

[4] *California's Ocean Resources: An Agenda for the Future* (Sacramento: California Resources Agency, March 1997), p. 5E-4.

[5] U. S. Department of the Interior, Minerals Management Service, *Outer Continental Shelf Draft Proposed Oil and Gas Leasing Program 1997 to 2002* (Washington, D. C., July 1995), p. x.

[6] U. S. Department of the Interior, Minerals Management Service, *Proposed Final Outer Continental Shelf Oil and Gas Leasing Program, 1997 to 2002* (Washington, D. C., August 1996), pp.1-4, 1-5.

[7] Memorandum to All Prospective Offerors from the Minerals Management Service, April 26, 1993. "Prospective Offerors" refers to persons or groups contemplating bidding for the study contract. The study was titled the *California Offshore Oil and Gas Energy Resources (COOGER)*.

[8] MMS Solicitation 3755 for proposals for *California Offshore Oil and Gas Energy Resources (COOGER)* study, Section C: Description/Specification/ Statement of Work (May 27, 1994) p. 3

[9] U. S. Department of the the Interior, Minerals Management Service, Pacific Region, *Santa Barbara County: Two Paths*, OCS Study MMS 96-0036, *San Luis Obispo County: A Major Switching*, OCS Study MMS 96-0037, and *Ventura County: Oil, Fruit, Commune, and Commute,* OCS Study MMS 96-0035 (Camarillo: Pacific OCS Region, Minerals Management Service).

[10] Some background is relevant to Mobil's Clearview project and the *ARCO v. California* dispute. The state Lands Commission in 1987 denied permission for ARCO to establish two additional offshore platforms off the UCSB campus. In settling ARCO's lawsuit against the state, ARCO quitclaimed two Ellwood offshore leases in exchange for permission to increase production off Long Beach. ARCO later sold its fifty percent share of the Ellwood lease rights to Mobil, its partner, which then sought to access the large offshore field from onshore. Opponents argued that the quitclaimed leases had become part of the adjacent sanctuary and therefore protected from further exploitation. Moreover, they said, directional drilling from onshore was environmentally and aesthetically unacceptable because it would simply replace offshore platforms with onshore industrialization. The dispute became moot, at least for the time, when Mobil abandoned its directional drilling plan.

Chapter 15

[1] Michael G. Powers, "Monitoring and Mitigating the Socioeconomic Impacts of Coastal Oil and Gas Development," *Proceedings of the Coastal Zone '91 Symposium on Coastal and Ocean Management* (Long Beach, July 8-12, 1991).

[2] Harvey Molotch and John Woolley, "Evaluation of Current Programs to Identify and Mitigate Socioeconomic Impacts in the Santa Barbara Channel: An Analysis of SEMP," May 17, 1994.

[3] Santa Barbara County Planning and Development Department report to Board of Supervisors for Oct. 14, 1994 hearing on 1994 and 1995 allocations from the Coastal Resources Enhancement Fund, p. 5.

[4] *California's Ocean Resources: An Agenda for the Future* (Sacramento: California Resources Agency, March 1997), p. 5E-12.

[5] "[A]s a result of its energy policies and programs, California used 36 percent less energy per dollar of economic output than the national average in 1993, an energy equivalent savings of approximately 687 million barrels of oil per year. In fact, California's per capita energy consumption was 19 percent lower in 1993 compared to 1975, while U. S. average per capita energy consumption was only 0.7 percent lower. On an energy equivalent basis, Californians use approximately 38 barrels of oil per year per capita, compared to a national average of 56 barrels." The quotation is from *California Ocean Resources: An Agenda for the Future* (p. 5E-5), issued in March 1997 by the California Resources Agency, which in turn attributes the information to 1995 reports of the California Energy Commission and the Office of Energy Markets and End Use.

Appendix A

Federal Oil And Gas Lease Sales Offshore California

Number	Date	Region[1]	No. Tracts	No. Bid On	Total High Bids
P-1	5-14-63	C-N	129	58	$12.8 million
P-3	12-15-66	S	1	1	$21.2 million
P-4	2-6-68	S	110	75	$602.7 million
35	12-11-75	S[2]	231	70	$417.3 million
48	6-29-79	S	148	55	$527.8 million
53	5-28-81	S	111	81	$2.088 billion
68	6-11-82	S	140	35	$117.9 million
73	11-30-83	S	173	8	$16.0 million
80	10-17-84	S-C	657	25	$62.1 million

[1] S (Southern), C-N (Central and Northern), S-C (Southern and Central)
[2] Sale 35 excluded the Santa Barbara Channel.

Appendix B

State of California Offshore Oil and Gas Leases

Parcel Number	Present Lessees or Last Lessees	Acreage	Date Issued
Santa Barbara County			
129	ARCO	254	1-27-44
208	ARCO	1,920	1-18-46
308	ARCO (quitclaimed 12-31-91)	1,920	3-04-47
309	ARCO (quitclaimed 12-31-91)	1,920	3-04-47
421	Venoco	68.5	10-22-29
424	ARCO (quitclaimed 11-3-71)	20	11-12-29
428	Wm. L. Appleford (quitclaimed 6-20-80)	69.2	7-29-30
1824	Chevron-Exxon	5,500	1-10-57
2198	SOCAL (quitclaimed)	3,840	10-14-58
2199	Chevron-Molino Energy	3,840	7-25-58
2205	Phillips-Pauley (quitclaimed 4-8-68)	3,840	7-25-58
2206	Texaco	3,840	7-25-58
2207	Phillips-Exxon et al (quitclaimed 7-25-75)	3,840	7-25-58
2725	Texaco	4,250	5-04-61
2726	ARCO et al	4,250	5-04-61
2793	ARCO et al	4,250	10-26-61
2879	Unocal	5,653	4-26-62
2894	Chevron-Molino Energy	4,250	6-28-62
2920	Chevron-Molino Energy	4,250	8-28-62
2933	Phillips-Pauley et al	4,250	10-25-62
2955	Phillips (quitclaimed 3-26-92)	4,250	10-20-62
2991	Unocal	4,250	2-28-63
3004	Unocal-Exxon	3,150	4-25-63

Parcel Number	Present Lessees Or Last Lessees	Acreage	Date Issued
3120	Venoco	3,324	4-29-64
3133	Exxon	5,535	5-28-64
3150	ARCO-Chevron	5,553	7-28-64
3242	Venoco	4,290	4-08-65
3498	Chevron-Exxon (quiitclaimed 2-5-92)	1,165	6-15-66
3499	Phillips-Pan Petroleum	1,340	6-15-66
3503	Unocal	1,660	6-28-66
4000	ARCO-Chevron	204	8-28-68
4001	Chevron (quitclaimed 3-12-90)	780	8-28-68
4002	Chevron (quitclaimed 11-25-91)	600	8-28-68
4031	Continental (quitclaimed 9-23-82)	2,332	8-28-68

Ventura County

Parcel Number	Present Lessees Or Last Lessees	Acreage	Date Issued
145	Energy Development Corp.	326	7-03-44
410	Berry	50	4-17-29
427	Mobil	1,485	5-19-30
429	Berry	80	4-21-31
735	Berry	220	6-30-52
1466	Berry	1,175	8-29-55
3184	Chevron (quitclaimed 12-17-93)	5,540	9-24-64
3314	Berry	5,430	7-02-65
3403	Chevron (quitclaimed 12-17-93)	5,300	11-18-65
3489	Mobil (quitclaimed 4-28-92)	4,570	5-26-66
3490	Mobil (quitclaimed 4-28-92)	5,305	5-26-66
3945	Shell (quitclaimed 5-19-93)	5,546	5-23-68
3946	Shell (quitclaimed 4-18-73)	5,589	5-23-68

Parcel Number	Present Lessees or Last Lessees	Acreage	Date Issued
Orange County			
91	Shell	589	5-21-43
163	Shell	640	11-15-44
186	Exxon-Texaco	1,255	9-24-45
E392	Shell	835	9-26-38
425	Shell	835	2-10-50
426	Shell	640	2-10-50
3033	Unocal	2,113	7-25-63
3095	Unocal	3,360	1-30-64
3119	Mobil (quitclaimed 4-25-73)	3,420	4-29-64
3177	Pauley Petroleum (quitclaimed 9-17-85)	1,549	8-28-64
3413	Unocal	1,871	12-01-65
Los Angeles County			
3455	ARCO Long Beach, Inc. (THUMS is contractor)	592	4-01-66

Appendix C

Major California and Federal Legislation, Case Law, Executive Orders and Administrative Actions Pertaining to Offshore Oil and Gas Leasing

1920 — Federal Mineral Leasing Act set regulations for leasing of federal lands for oil production. Its reach did not extend beyond the water's edge.

1921 — California Mineral Leasing Act authorized issuing offshore oil and gas prospecting permits. Waters fronting on cities were protected.

1928 — California Supreme Court in *Boone v. Kingsbury* upheld the validity of the 1921 act and ordered the state to approve applications for offshore prospecting and drilling permits.

1928 — Legislature declared emergency moratorium on offshore leases. Previously-leased tracts were not affected.

1929 — California legislature closed tidelands to further leasing.

1931 — Legislature lifted the ban on additional offshore leasing. "This measure was defeated by referendum. Two initiative measures, one for tidelands drilling and one for slant-well operation, were also defeated."[1]

1938 — State Lands Act created the State Lands Commission with authority to lease state lands for oil production, and to regulate such operations. Wells could be drilled into the tidelands only from onshore or from filled land and then only to protect tidelands that were threatened from drainage by onshore wells.

1945 — (September) President Truman proclaimed that all waters off the nation's coasts, up to the shoreline, were under federal jurisdiction.

1946 — (July) Congress adopted a resolution quitclaiming the tidelands to the states.

1946 — (August) President Truman vetoed the quitclaim resolution.

1947 — U. S. Supreme Court, in *United States v. California*, ruled that the federal government, not California, owned the waters extending three miles offshore.

1952 — (May 16) Congress passed legislation quitclaiming tidelands to the states.

1952 — (May 29) President Truman vetoed the quitclaim bill.

1953 — Congress passed the Federal Submerged Lands Act which placed California tidelands, extending three miles seaward, under state jurisdiction.

1953 — Congress passed the Federal Outer Continental Shelf Lands Act which delegated to the Interior Department authority to lease ocean parcels beyond the three-mile line for oil production, and to regulate that activity.

1955 — State legislature passed California Cunningham-Shell Tidelands Act. Oil companies were now authorized to install offshore drilling platforms. It also established sanctuaries in which no leasing would be permitted.

1964 — State Senate Bill 60 granted the eastern portion of the Wilmington Oil Field on tidelands to the City of Long Beach. This led to the city's leasing to THUMS (Texaco, Humble, Union, Mobil and Shell) which built production islands in the Long Beach Harbor.

1965 — U. S. Supreme Court ruled in *U. S. v. California* that the entire Santa Barbara Channel is *not* "inland waters," and therefore not totally under state jurisdiction. Only waters extending three miles from the mainland and around the channel islands are tidelands under state control.

1967 — The Interior Department created a two-mile wide ecological preserve seaward of the state sanctuary off Santa Barbara in which no oil leases would be offered.

1969 — Interior Secretary Walter Hickel established new regulations for oil activity in the Santa Barbara Channel and created an additional buffer zone seaward of the federal preserve established in 1967.

1970 — The National Environmental Protection Act (NEPA) was signed by President Nixon. It required, among other things, that the federal government prepare impact reports on offshore oil leasing plans. It also created the President's Council on Environmental Quality.

1971 — Governor Reagan signed State Senator Robert Lagomarsino's bill creating an oil-free sanctuary extending three miles seaward around the Santa Barbara Channel Islands.

1972 — California voters approved Proposition 20, the California Coastal Act which established the Coastal Commission and which led eventually to disputed authority over state offshore oil leasing and equally-disputed authority over federally-permitted oil activity in the Outer Continental Shelf.

1972 — The federal Coastal Zone Management Act was passed which granted to coastal states limited oil leasing authority in the federal Outer Continental Shelf.

1975 — The Outer Continental Shelf Advisory Board was established to receive comments from state officials, federal agencies, and other interested parties on offshore oil leasing and operating policies.

1978 — Congress significantly amended the Outer Continental Shelf Lands Act of 1953. The amendment required the Interior Department to prepare five-year leasing plans with the participation of adjacent states and coastal communities. The commercial fishing industry was also offered some protection. An oil spill liability fund was established to pay for cleaning up spills and related damages. The Bureau of Land Management was made the leasing agency, and the Geological Survey was responsible for supervising offshore oil operations.

1980 — President Carter signed the Channel Islands National Marine Sanctuary designation. Implementing regulations were suspended for a year by the succeeding Reagan administration, but took effect in 1981.

1980 — Interior Secretary Cecil Andrus released the first five-year (1980-1985) leasing plan.

1981 — The U. S. Court of Appeals sent the 1980-85 five-year plan crafted by Cecil Andrus, former Interior secretary, back to the Interior Department for greater elaboration on environmental and economic impacts.

1981 — The U. S. Commerce Department, which administers the Coastal Zone Management Act, revised regulations to specifically deny the California Coastal Commission any voice in federal OCS leasing. New regulations maintained that leasing itself does not directly affect coastal resources. Under congressional pressure, the department withdrew the new rule.

1982 — The Interior Department established the Minerals Management Service to assume the leasing and supervising authority previously shared by the Bureau of Land Management and the Geological Survey.

1982 — Interior Secretary James Watt released the second five-year (1982-1987) leasing plan.

1982 — Congress imposed the first of a series of annual moratoria on federal offshore oil leasing off California.

1984 — The U. S. Supreme Court ruled that under the federal Coastal Zone Management Act the California Coastal Commission did not have authority to delay a federal lease sale because the sale itself has no direct affect on the state coastal resources. The 5-4 decision overturned the rulings of the District Court and the Court of Appeals.

1987 — Interior Secretary Donald Hodel released the third five-year (1987-1992) leasing plan.

1988 — The U. S. Court of Appeals upheld the 1987-1992 five-year leasing plan against most legal challenges, but ruled that it contained inadequate information on potential cumulative impacts on migratory species.

1990 — President Bush cancelled three pending California offshore lease sales. There would be no more California lease sales before the turn of the century, except possibly off Santa Barbara and San Luis Obispo counties where leasing might be resumed after 1996.

1990 — Congress dropped the requirement that federal offshore lease sales must "directly" affect state coastal resources to be subject to state review and objection. The action negated the U. S. Supreme Court ruling of 1984 and restored state authority over lease sales under the federal Coastal Zone Management Act.

1990 — Congress amended the 1970 Clean Air Act to transfer air quality regulatory authority on the Outer Continental Shelf from the Interior Department to the Environmental Protection Agency which in turn was authorized to delegate that authority to local jurisdictions, including Santa Barbara County.

1991 — The Supreme Court in January ruled that California cities and counties should be allowed to restrict onshore facilities used in support of offshore oil and gas drilling and production. The oil industry had challenged laws in five counties and five cities that restrict or ban such onshore facilities. Some local laws were enacted by initiatives. See Appendix D.

1994 — Governor Pete Wilson signed a bill drafted by Assemblyman Jack O'Connell which banned any further oil leasing in the state tidelands. It did not interfere with previously-awarded leases.

1995 — The Interior Department solicited public comments on the fourth five-year (1997-2002) federal offshore leasing plan.

1996 — The Interior Department exempted California from the federal five-year (1997-2002) offshore leasing plan.

[1] Ernest R. Bartley, *The Tidelands Oil Controversy: A Legal and Historical Analysis* (Austin: University of Texas Press, 1953), p.71. He quotes testimony of Thomas A. Dockweiler, a former congressman from California, during congressional hearings on federal quitclaims legislation in 1939.

Appendix D

Actions of California Coastal Cities and Counties Against Offshore Oil Development, 1985 Through 1996

Communities	Date	Means of Passage Election	Means of Passage Board/Council	Type
Cities				
Santa Cruz	11/85	82 percent		A
San Luis Obispo	6/86	78 percent		A
San Diego	11/86	77 percent		P
Oceanside	11/86	64 percent		P
Morro Bay	11/86	70 percent		P
Monterey	11/86	76 percent		A
San Francisco	11/86	71 percent		M
Redondo Beach	1/87		5-0	A
Point Arena	2/87		3-0	A
Capitola	4/87		5-0	P
Pacific Grove	5/87		7-0	P
Half Moon Bay	11/87	69 percent		P
San Clemente	11/88	75 percent		A
Fort Bragg	11/88		4-1	A
Trinidad	12/88		4-1	A
Carlsbad	9/89		5-0	P
Watsonville	9/90		7-0	A
Laguna Beach	11/90	85 percent		A
Counties				
San Francisco	8/90		6-1	R
Santa Cruz	6/86	78 percent		A
Sonoma	11/86	74 percent		A
San Mateo	11/86	61 percent		A
Monterey	11/86	74 percent		A
San Luis Obispo	11/86	53 percent		A
San Diego	11/86		5-0	A
Mendocino	11/88	79 percent		A
Humboldt	11/88	68 percent		A
Santa Barbara	3/96	54 percent		B

A—Policy/land use changes for oil facilities require vote.
B—Prohibition on oil facilities without voter-approval.
M—Moratorium for two years.
R—Restrict oil facilities to "overlay district" with conditonal use.

Sources: Save Our Shores, Santa Cruz.
1996 Santa Barbara County vote, Santa Barbara County Elections Office.

Appendix E*

Oil And Gas Production Platforms Off California

Outer Continental Shelf

Name: HOGAN
Operator: Formerly Phillips Petroleum, now Pacific Operators
Location: Carpinteria Field 3.7 miles offshore
Water depth: 150 feet
Year installed: 1967
Number of well slots: 66

Name: HOUCHIN
Operator: Formerly Phillips Petroleum, now Pacific Operators
Location: Carpinteria Field 4.1 miles offshore
Water Depth: 163 feet
Year installed: 1968
Number of well slots: 60

Name: PLATFORM A
Operator: Formerly Union Oil (Unocal), now Torch
Location: Dos Cuadros Field, 5.8 miles offshore Santa Barbara
Water depth: 190 feet
Year installed: 1968
Number of well slots: 57

Name: PLATFORM B
Operator: Formerly Union Oil (Unocal), now Torch
Location: Dos Cuadros Field, 5.7 miles offshore Santa Barbara
Water depth: 190 feet
Year installed: 1968
Number of well slots: 63

Name: HILLHOUSE
Operator: Originally Sun Oil, now operated by Torch.
Location: Dos Cuadros Field, 5.5 miles offshore Santa Barbara
Water depth: 190 feet.
Year installed: 1969
Number of well slots: 60

Name: HONDO
Operator: Exxon
Location: Santa Ynez Unit, 5.1 miles offshore Gaviota
Water depth: 850 feet
Year installed: 1976
Number of well slots: 28

Name: PLATFORM C
Operator: Formerly Union Oil (Unocal), now operated by Torch
Location: Dos Cuadros Field, 5.7 miles offshore Santa Barbara
Water depth: 190 feet
Year installed: 1977
Number of well slots: 60

Name: GRACE
Operator: Chevron
Location: Santa Clara Unit, 10.5 miles offshore Ventura/Oxnard
Water depth: 318 feet
Year installed: 1979
Number of well slots: 48

Name: HENRY
Operator: Formerly Sun Oil (Unocal), now operated by Torch
Location: Carpinteria Field, 4.3 miles off Carpinteria
Water depth: 173 feet
Year installed: 1979
Number of well slots: 24

Name: ELLEN
Operator: Formerly Shell Western Exploration and Production, now operated by Cal Resources.
Location: Beta Unit, 8.6 miles off Huntington Beach
Water depth: 265 feet
Year installed: 1980
Number of well slots: 80

Name: ELLY
Operator: Formerly Shell Western Exploration and Production, now operated by Cal Resources.
Location: Beta Unit, 8.6 miles offshore Huntington Beach
Water depth: 255 feet
Year installed: 1980

Name: GINA
Operator: Formerly Unocal, now operated by Torch
Location: Hueneme Field, 3.7 miles offshore Port Hueneme
Water depth: 95 feet
Year installed: 1980
Number of well slots: 15

Name: GILDA
Operator: Formerly Unocal, now operated by Torch
Location: Santa Clara Field 3,7 miles offshore Oxnard
Water depth: 95 feet
Year installed: 1980
Number of well slots: 96

Name: HABITAT
Operator: Texaco
Location: Pitas Point Field, 7.8 miles offshore Carpinteria
Water depth: 290 feet
Year installed: 1981
Number of well slots: 24

Name: EDITH
Operator: Formerly Unocal, now operated by Torch
Location: Beta Field, 8.5 miles offshore Huntington Beach
Water depth: 161 feet
Year installed: 1983
Number of well slots: 72

Name: EUREKA
Operator: Formerly Shell Western Exploration and Production, now operated by Cal Resources
Location: Beta Field, 9 miles offshore Huntington Beach
Water depth: 700 feet
Year installed: 1984
Number of well slots: 60

Name: HARVEST
Operator: Formerly Texaco, now Chevron
Location: Point Arguello Field, 6.7 miles offshore Point Conception/ Point Arguello
Water depth: 675 feet
Year installed: 1985
Number of well slots: 50

Name: HERMOSA
Operator: Chevron
Location: Point Arguello Field, 6.8 miles offshore Point Conception/
Point Arguello
Water depth: 603 feet
Year installed: 1985
Number of well slots: 48

Name: IRENE
Operator: Formerly Unocal, now Torch
Location: Point Pedernales Field, 4.7 miles offshore Point Arguello
Water depth: 242 feet
Year installed: 1985
Number of well slots: 72

Name: HIDALGO
Operator: Chevron
Location: Point Arguello Field, 5.9 miles offshore Point Arguello/
Point Conception.
Water depth: 430 feet
Year installed: 1986
Number of well slots: 56
Name: GAIL
Operator: Chevron
Location: Sockeye Field, 9.9 miles offshore Oxnard/Port
Hueneme
Water depth: 739 feet
Year installed: 1987
Number of well slots: 36

Name: HARMONY
Operator: Exxon
Location: Santa Ynez Unit 6.4 miles offshore Gaviota
Water depth: 1,198 feet
Year installed: 1989
Number of well slots: 60

Name: HERITAGE
Operator: Exxon
Location: Santa Ynez Unit 8.2 miles offshore Gaviota
Water depth: 1,075 feet
Year installed: 1989
Number of well slots: 60

* Listed in chronological order of installation

Appendix F*

Oil And Gas Production Islands and Platforms Off California

State Tidelands

Name: MONTEREY ISLAND
Owners and operators: Originally Monterey Oil Co. and The Texas Co. (Texaco); later purchased by Exxon
Location: Belmont Field off Seal Beach
Water depth: 42 feet
Year installed: 1954
Number of well slots: 70

Name: RINCON ISLAND
Owners and operators: Built by Atlantic Richfield Co. (ARCO), later sold to Norris Oil Co.
Location: Rincon Field, Ventura County.
Water depth: 45 feet
Year installed: 1958
Number of well slots: 46

Name: HAZEL
Owner-operator: Standard Oil of California (Chevron)
Location: Summerland Field
Water Depth: 100 feet
Year Installed: 1958
Year Removed: 1996
Number of well slots: 25

Name: HILDA
Owner-operator: Standard Oil of California (Chevron)
Location: Summerland Field
Water Depth: 106 feet
Year Installed: 1960
Year Removed: 1996
Number of well slots: 24

Name: HELEN
Owner-operator: Texaco
Location: Cuarta Field (Western Santa Barbara Channel)
Water Depth: 94 feet
Year Installed: 1960
Year Removed: 1988
Number of well slots: 40

Name: HARRY
Owner-operator: Phillips Petroleum
Location: Conception Offshore (Western Santa Barbara Channel)
Water Depth: 105 feet
Year Installed: 1961
Year Removed: 1974
Number of well slots: 45

Name: HERMAN
Owner-operator: Texaco
Location: Conception Offshore
Water Depth: 85 feet
Year Installed: 1963
Year Removed: 1988
Number of well slots: 20

Name: EMMY
Owner-operator: Aminoil, later Shell-Western Exploration & Production (SWEPI)
Location: Huntington Beach
Water Depth: 47 feet
Year Installed: 1963
Number of well slots: 53

Name: EVA
Owner-operator: Unocal
Location: Huntington Beach
Water depth: 57 feet
Year installed: 1964
Number of well slots: 39

Name: ESTHER
Owner-operator: Unocal
Location: Belmont Field off Seal Beach
Water depth: 38 feet
Year installed: 1965 as an island, later rebuilt at platform.

Name: THUMS islands (4)
Owner: ARCO Long Beach
Operator: THUMS (Texaco, Humble [now Exxon], Union [now Unocal], Mobil, and Shell). THUMS, former owner, now a subsidiary of ARCO Long Beach.
Location: Wilmington Field, in Long Beach Harbor
Water depth: 25 to 40 feet
Year installed: 1965

Name: HEIDI
Owner-operator: Standard Oil of California (Chevron)
Location: Carpinteria Field
Water Depth: 125 feet
Year installed: 1966
Year removed: 1996
Number of well slots: 60

Name: HOPE
Owner-operator: Standard Oil of California (Chevron)
Location: Carpinteria Field
Water Depth: 140 feet
Year installed: 1966
Year removed: 1996

Name: HOLLY
Owner-operator: Previously ARCO and Mobil; now Venoco
Location: South Ellwood Field
Water depth: 211 feet
Year installed: 1966
Number of well slots: 30

* Listed in chronological order of completion or installation

Appendix G

California Governors, Lands Commission Executive Officers and Coastal Commission Executive Directors

California Governors Since 1943

Earl Warren, 1943-1955
Goodwin Knight, 1955-1959
Edmund G. (Pat) Brown, 1959-1967
Ronald Reagan, 1967-1975
Edmund G. (Jerry) Brown Jr., 1975-1983
George Deukmejian, 1983-1991
Pete Wilson, 1991-

Executive Officers, California State Lands Commission, since 1957

Francis Hortig, 1957-1972
Richard Golden, 1972-1973
Ed Gladish, 1973-1975
William Northrup, 1975-1982
Claire Dedrick, 1982-1990
Charles Warren, 1990-1994
Robert Hight, 1994-

Executive Directors, California Coastal Commission, since 1973

Joseph Bodovitz, 1973-1978
Michael Fischer, 1978-1985
Peter Douglas, 1985-

Appendix H

Department of the Interior Secretaries and Directors of the Minerals Management Service and Geological Survey

Interior Secretaries Since 1933

Harold Ickes	1933-1946	Presidents Roosevelt and Truman
Julius Krug	1946-1949	President Truman
Oscar Chapman	1949-1953	President Truman
Douglas McKay	1953-1956	President Eisenhower
Fred A. Seaton	1956-1961	President Eisenhower
Stewart Udall	1961-1969	Presidents Kennedy and Johnson
Walter Hickel	1969-1970	President Nixon
Rogers Morton	1971-1975	Presidents Nixon and Ford
Stanley Hathaway	1975-1975	President Ford
Thomas Kleppe	1975-1977	President Ford
Cecil Andrus	1977-1981	President Carter
James Watt	1981-1983	President Reagan
William Clark	1983-1985	President Reagan
Donald Hodel	1985-1989	President Reagan
Manuel Lujan Jr.	1989-1993	President Bush
Bruce Babbitt	1993-	President Clinton

Directors, Minerals Management Service, Since 1982

Harold Doley	1982-1983
William Bettenberg	1983-1988
Robert Kallman	1988-1989
Barry A. Williamson	1989-1991
S. Scott Sewell	1991-1993
Tom Fry	1993-1994
Cynthia Quarterman	1994-

Directors, Geological Survey, Since 1933

Walter Mendenhall	1930-1943	Vincent McKelvey	1971-1978
William Wrather	1943-1956	Henry Menard	1978-1981
Thomas Nolan	1956-1965	Dallas Peck	1981-1993
William Pecora	1965-1971	Gordon Eaton	1994-

Appendix I

Glossary and Abbreviations

APCD: Air Pollution Control District, usually a county-level agency responsible for collecting air quality control information and advising its governing body on ways to meet state and federal clean air standards.

Asphalt: Thick or solid hydrocarbons found in some crude oil that is more suitable for producing paving material than gasoline or other light petroleum products.

Army Corps of Engineers: Agency that oversees or conducts engineering and navigational safety activities on the nation's navigable waters, including the Outer Continental Shelf lands.

Barite: Mineral used as a weighting material in drilling mud.

Barrel: A measure of petroleum products equal to 42 gallons. It is an accepted statistical fiction, as barrels used in the oil industry contain 55 gallons.

Basin: Geological formation under the surface, called a syncline, which is a likely location for petroleum deposits.

Bitumen: Solid or semi-solid hydrocarbons. (See Asphalt)

Bituminous: Containing bitumen.

BLM: Bureau of Land Management, an Interior Department agency that at one time conducted offshore oil lease sales for the federal government. That responsibility was assumed in 1982 by the Minerals Management Service.

Blowout: Uncontrolled eruption of oil and/or gas up through a well head due to loss of pressure control in the well.

Board of Supervisors: Governing body of a California county, usually consisting of five elected representatives serving four-year terms.

Buoy: An anchored float placed to indicate vessel channels, navigational hazards, fishing gear, etc.

California Coastal Commission: Reviews permits for coastal projects, including tidelands oil activities. It has authority under the Coastal Zone

Management Act to review and pass on oil development, including lease sales, on the Outer Continental Shelf.

California State Lands Commission: A three-member group that manages state lands, conducts state offshore lease sales and regulates offshore oil activities.

Casing: The outer steel pipe, or wall, of an oil and/or gas well.

CEQ: Council on Environmental Quality, a White House advisory panel.

CEQA: California Environmental Quality Act. Mandates environmental studies, hearings, and disclosure of information for developments requiring state permits.

COLAB: Coalition of Labor, Agriculture, and Business. A nongovernmental group in Santa Barbara County representing conservative business interests and in which labor plays almost no role.

Condensate: Liquid hydrocarbons produced with natural gas.

Core drilling: Drilling small-bore holes in the earth and extracting the material drilled to determine the nature of the subsurface.

CZMA: Coastal Zone Management Act of 1972 provided for federal grants for development and protection of coastal resources to states that had prepared federally-approved coastal protection plans. It also conveyed some authority to states over federally-regulated development activities in the Outer Continental Shelf.

Derrick: A wood or steel frame tower over an oil or gas well that supports drilling equipment and from which drill pipe and casing is lowered into the well.

Dispersant: A chemical that reduces water surface tension, breaking up a slick and enabling oil to sink or enter the water column. It is sprayed over spills to prevent oil from drifting to areas targeted for protection.

Distillate: See condensate.

Drainage: Migration of oil in subsurface deposits due to changes in pressure as result of extraction of oil, gas, and/or water through wells.

Drill cuttings: Rock fragments and other debris loosened by the oil well drill bit and brought to the surface.

Drill pipe: Steel pipe that rotates and turns the drill bit at the bottom end. Drilling mud is circulated down through the pipe and back up between the pipe and the outer wall of the well hole. Drill pipe is in sections about 30 feet long, and sections are added as the well is drilled deeper.

Drilling mud: A mixture of clay, water, and additives circulated through an oil or gas well. It has several functions. It controls pressure in the well, it cools the drill bit, it carries rock cuttings to the surface, and it seals and stabilizes the wall of the well hole.

Drilling rig: The equipment used in drilling oil and gas wells. It includes the derrick and the power source to drive the drill pipe and the drilling bit.

EPA: Environmental Protection Agency. There is such an agency in both the federal and California governments. EPA establishes and enforces standards for air and water quality, waste disposal, waste cleanup, and other environmental activities.

EIR: Environmental Impact Report, a document required by the state of California for developments likely to have a significant impact on natural resources.

EIS: Environmental Impact Statement, a document required by the federal government for developments likely to have a significant impact on natural resources.

EQAB: Environmental Quality Advisory Board, a municipal group established by the city of Santa Barbara after the 1969 offshore oil well blowouts.

Exploratory drilling: Drilling to determine whether sufficient oil exists at a site to proceed with commercial production. It also can determine the areal limits of an oil deposit.

Geophysical (or seismic) surveys: Shock wave and other techniques for determining subsurface geological structures. They do not find oil deposits, but they indicate structures in which such deposits are most likely to exist.

FERC: Federal Energy Regulatory Commission, the agency that reviews and passes on energy projects requiring federal permits.

GOO: Get Oil Out, an organization founded immediately after the 1969 oil well blowouts off Santa Barbara. Its goal has been to eliminate offshore oil production in the Santa Barbara area.

Gravity: Ratio of the density of a liquid to the density of water. Oil rated 28 to 29 degrees gravity is light and easy to refine into gasoline. Heavy oil, rated 10 to 20 degrees gravity, is more likely to become asphalt for road paving.

Hydrogen sulfide: A poisonous, flammable gas that smells like rotten eggs and is found in "sour" natural gas.

Impact report: See EIR.

Interior: The Department of the Interior is the federal agency responsible for overseeing oil and gas operations in the Outer Continental Shelf.

Jackup rig: An offshore oil drilling vessel with three or four racheted legs that are extended to the ocean floor when in operation, and jacked up when being towed.

Leasing: Granting contracts for exploring for and producing oil and/or gas in return for a share of the production earnings. Offshore oil and gas leases are ordinarily awarded by states or the federal governments in competitive bidding.

Liquified natural gas (LNG): Natural gas reduced to .06 of its normal volume by dropping its temperature to minus 260 degrees Fahrenheit. In this liquefied state, it can be shipped in ultra-insulated tankers. Extremely toxic and flammable, it is devastating in the event of a leak.

Liquefied petroleum gas (LPG): A mixture of petroleum gases that become liquid under pressure. Butane and propane are the principle kinds of LPGs.

MMS: Minerals Management Service. Interior Department agency created in 1982 to assume, among other duties, administration of offshore oil and gas leasing and regulatory operations under the leases. These functions had previously been carried out by the Bureau of Land Management and the Geological Survey.

Moonpool: Open vertical hole amidships in a drilling vessel through which offshore oil wells are drilled.

Mudline: The bottom of an ocean or lake.

NAS/NRC: National Academy of Sciences/National Research Council. The NRC is an independent advisor to the federal government on scientific and technical questions. It is part of the NAS. Both operate under a congressional charter, but are private organizations with no congressional funding.

NEPA: National Environmental Policy Act, signed into law Jan. 1, 1970, requires a study of potential environmental consequences of major undertakings before federal permits may be issued.

NIMBY: Not In My Back Yard. NIMBIES are people and groups who protest the location of facilities in their neighborhoods because of noise, pollution, congestion, health and safety risks, etc.

NMFS: National Marine Fisheries Service, a U. S. Commerce Department agency that conducts or sponsors research on the nation's fisheries, advises the national government on fisheries management, and prepares fisheries management regulations.

NOAA: National Oceanic and Atmospheric Administration, a U. S. Commerce Department agency which conducts oceanic and atmospheric research, provides weather information, and participates in coastal resource management, among other duties.

OCS: Outer Continental Shelf. The part of the offshore continental shelf beyond the first three miles which are under state jurisdiction. The OCS is federal territory.

OCSLA: Outer Continental Shelf Lands Act of 1953, substantially amended in 1978. It authorizes the Interior Department to grant leases for oil and gas exploration and production on the OCS. It also establishes federal regulatory authority and responsibility in this area.

Offloading: Discharging (unloading) fluid cargo from a tanker.

OPEC: Organization of Petroleum Exporting Countries, approximately a dozen Third World nations, about half of which are in the Middle East.

PIRO: Petroleum Industry Response Organization, formed by the American Petroleum Institute to cope as promptly and effectively as possible with major oil spills.

PCB: Polychlorinated biphenyl, once used in oil and gas processing equipment as a heat transfer medium, but later banned by the U. S. Environmental Protection Agency as a carcinogen.

PG&E: Pacific Gas and Electric Company, a major private, publicly-regulated energy utility operating in Central and Northern California.

Platform: Offshore structure secured to the ocean bottom, with above-water working decks from which exploration and production oil and gas wells are drilled. A platform has slots for many wells.

PUC: Public Utilities Commission, used here to designate the California agency that reviews and passes on energy utilities' projects that require state approval.

Quitclaim: The release of contractual (including lease) interest in land. The interest reverts back to the lessor.

Refinery: A plant where petroleum liquids are made into fuels, lubricants, asphalt or other products or feedstocks for other industrial processes. A refinery is distinguished from a processing plant which handles only preliminary treatment of crude, such as separating water produced with the oil.

Rig: A drilling rig, including derrick and machinery to turn the drill pipe and drill bit.

Royalty: A percentage of the value of oil produced that is paid by the producer to the owner or lessor of the land.

Seeps: Oil and tar that emerges through the surface of the earth (including the ocean bottom) from shallow deposits that have little or no caprock to contain the petroleum materials.

Seismic testing, seismic surveying: Investigating underground geologic formations by recording and analyzing shock waves artificially produced and reflected from the subsurface.

Semisubmersible drilling vessel: Marine oil drilling vessel which is stabilized in the water by flooding large tanks to add weight and make the vessel float very low in the water. Ordinarily it does not have propulsion, and is moved by tugs when the stabilizing tanks, or hulls, are pumped out and the vessel floats high in the water.

SID: Secretarial Information Document prepared by the Interior Department staff for the secretary who then determines, for example, whether an offshore lease sale should be conducted and under what terms.

Skimmer: Any one of a number of devices designed to remove spilled oil from the surface of the water. It is usually part of a floating vessel, although some are operated from land.

Slant drilling: This is also called directional drilling or whipstocking. It is a technology for aiming the drill bit and drill pipe in a direction other than straight down.

Subsea Completion: An oil or gas well with the wellhead on the oceanbottom and out of sight. It is used mostly in shallow water, and is remotely controlled.

Supervisors: See Board of Supervisors.

Tankering: Transporting oil or other petroleum products by marine tankers.

Tar: Non-fluid, sticky petroleum that is also called asphaltic or bituminous.

Tidelands: As used in the oil industry, they are state waters extending three miles seaward, and not simply the intertidal shoreline.

Tricounties: As used here, the Central California coastal counties of San Luis Obispo, Santa Barbara, and Ventura.

UCSB: University of California at Santa Barbara.

Unitization: An arrangement in which two or more contiguous oil leases are operated as a single production unit. One company is designated as operator. Unitization reduces the number of platforms required to produce oil from a number of leases.

USFWS: United States Fish and Wildlife Service. Interior Department agency which has research, advisory, and enforcement responsibilities in protecting fish and wildlife.

USGS: United States Geological Survey. The agency supervised offshore leasing in the OCS before that function was transferred to the Minerals Management Service in 1982. The USGS is primarily a scientific research and advisory agency.

Whipstocking: See slant drilling.

WOGA: Western Oil and Gas Association, trade group of oil companies in the western states. It was succeeded by the Western States Petroleum Association (WSPA).

Select Bibliography

Bartley, Ernest R. *The Tidelands Oil Controversy: A Legal and Historical Analysis.* Austin, University of Texas Press,1953.

Boesch, Donald F., Carl H. Hershner, and Jerome H. Milgram. *Oil Spills and the Marine Environment: Papers Prepared For the Energy Policy Project of the Ford Foundation.* Cambridge, Ballinger, 1974.

Davis, Lee Niedringhaus. *Frozen Fire: Where Will It Happen Next?* San Francisco, Friends of the Earth, 1979.

Dye, Lee. *Blowout at Platform A: The Crisis that Awakened a Nation.* New York, Doubleday, 1971.

Easton, Robert. *Black Tide: The Santa Barbara Oil Spill and Its Consequences.* New York, Delacourte, 1972.

ETA Offshore Seminars. *The Technology of Offshore Drilling, Completion and Production.* Tulsa, Petroleum Publishing Company, 1976.

Franks, Kenny A. and Paul Lambert, eds. *Early California Oil: A Photographic History, 1865-1940.* College Station, Texas A & M Press, 1985.

Freudenburg, William R. and Robert Gramling. *Oil In Troubled Waters: Perceptions, Politics, and the Battle Over Offshore Drilling.* Albany, State University of New York Press, 1994.

Gramling, Robert. *Oil on the Edge: Offshore Development, Conflict, Gridlock.* Albany, State University of New York Press, 1996.

Graves, Gregory R. and Sally L. Simon, eds. *A History of Environmental Review in Santa Barbara County, California.* Santa Barbara, University of California, 1980.

Harris, L. M. *Deepwater Floating Drilling Operations: An Introduction.* Tulsa, Petroleum Publishing Company, 1972.

Hershman, Marc J. and David L. Fluharty, and Scott L. Powell. *State and Local Influence Over Offshore Oil Decisions.* Seattle, University of Washington, 1988.

Hickel, Walter. *Who Owns America?* Englewood Cliffs, Prentice-Hall, 1971.

Holing, Dwight. *Coastal Alert: Ecosystems, Energy, and Offshore Oil Drilling.* Washington, D. C., Island Press, 1990.

Hoult, David P., ed. *Oil on the Sea.* New York, Plenum, 1969.

Johnson, G. Wesley and Ronald L. Nye, eds. *Environmental Hazards and Community Response: The Santa Barbara Experience.* Santa Barbara, University of California, 1979.

Jones, Charles S. *From the Rio Grande to the Arctic: The Story ofthe Richfield Oil Corporation.* Norman, University of Oklahoma Press, 1972.

Kallman, Robert E. and Eugene D. Wheeler. *Coastal Crude in a Sea of Conflict.* San Luis Obispo, Blake,1984.

Lambert, May. *Growing Up With Summerland, 1874-1975.* Carpinteria, Carpinteria Valley Historical Society, 1975.

Lima, James T. *The Politics of Offshore Energy Development* (dissertation), Santa Barbara, University of California, 1994.

Marx, Wesley. *Oilspill.* San Francisco, Sierra Club, 1971.

Mead, Walter J., Asbjorn Moseidjord, Dennis D. Muraoka, and Philip E. Sorensen. *Offshore Lands: Oil and Gas Leasing and Conservation on the Outer Continental Shelf.* San Francisco, Pacific Institute for Public Policy Research, 1985.

Nalder, Eric. *Tankers Full of Trouble: The Perilous Journey of Alaskan Crude.* New York, Grove, 1994.

Nash, A. E., Dean E. Mann and Phil G. Olsen. *Oil Pollution and the Public Interest: A Study of the Santa Barbara Oil Spill.* Berkeley, UC Institute of Governmental Studies, 1972.

Rintoul, William. *Drilling Ahead: Tapping California's Richest Oil Fields.* Santa Cruz, Valley Publishers, 1981.

_____. *Drilling Though Time: 75 Years with California Division of Oil and Gas.* Sacramento, California Division of Oil and Gas, 1990.

_____.*Spudding In.* San Francisco, California Historical Society, 1976.

Schuyler, Arent Henry Jr. *The Risks of Marine Traffic and Oil Operations in the Santa Barbara Channel and the Santa Maria Basin.* (dissertation) Los Angeles, University of California, 1996.

Steinhart, John and Carol. *Blowout: A Case Study of the Santa Barbara Oil Spill.* Duxbury Press, 1972.

Udall, Stewart, Charles Conconi, and David Osterhout. *The Energy Balloon.* New York, McGraw Hill, 1974.

Van der Linde, Peter, with Naomi A. Hintze. *Time Bomb: LNG: The Truth About Our Newest and Most Dangerous Energy Source.* Garden City, Doubleday, 1978.

Watkins, T. H. *Righteous Pilgrim: The Life and Times of Harold L. Ickes, 1874-1952.* New York, Henry Holt, 1990.

Wilder, Robert J. *Listening to the Sea: An Evolution Toward Ecological Ocean Governance* (forthcoming).

Public Documents

(Hundreds of public documents were consulted in the preparation of this book. A few of the most relevant are listed.)

California Resources Agency. *California's Ocean Resources: An Agenda for the Future.* Sacramento, 1997.

National Research Council/National Academy of Sciences. *Assessment of the U. S. Outer Continental Shelf Environmental Studies Program: III, Social and Economic Studies.* Washington, D. C., National Academy Press, 1992.

_____. *The Adequacy of Environmental Information for Outer Continental Shelf Oil and Gas Decisions: Florida and California.* Washington D. C., National Academy Press, 1989.

United States Geological Survey. *Geology, Petroleum Development, and Seismicity of the Santa Barbara Channel Region, California.* Geological Survey *Professional Paper* 679 (Washington, D. C.,1969).

United States Minerals Management Service. *Ventura County: Oil, Fruit, Commune, and Commute.* OCS Study MMS 96-0035. Camarillo, Pacific OCS Region, 1996.

_____. *Santa Barbara County: Two Paths.* OCS Study MMS 96-0036. Camarillo, Pacific OCS Region, 1996.

_____. *San Luis Obispo County: A Major Switching.* OCS Study MMS 96-0037. Camarillo, Pacific OCS Region, 1996.

_____. *Trends in Public Opinion on Offshore Oil Development in California.* By R. A. N. Smith, under contract with Minerals Management Service and Southern California Educational Initiative, 1995.

Chevron's Platform Grace 10.5 miles offshore Ventura and Oxnard in the Santa Barbara Channel. (Robert Sollen photo)

Oil platforms in a row about 5.5 miles offshore Santa Barbara in the Santa Barbara Channel. (Robert Sollen photo)

Index